THE ALMSHOUSES C

Front cover: Wandesford House, Bootham, York
Back cover: Ingram's Hospital, 1632-1957, Bootham, York

Fothergil Homes, Avenue Road, Clifton
The range built after the closure of Ingram's Hospital in 1957

THE

ALMSHOUSES

OF YORK

Medieval Charity to Modern Welfare

by

Carole Smith

Quacks Books, Petergate, York
in association with
the Sessions Book Trust
York England

ISBN 978-1-904446-30-9

British Library Cataloguing in Publication Data
Carole Smith 2010
The Almshouses of York

Printed and Published by
Quacks Books
7 Grape Lane Petergate
York YO1 7HU

Contents

Illustrations

The illustrations are credited in the text.
Photographs are by the author unless otherwise stated.

Maps

Drawn by Richard Parkin

Acknowledgements

There are many people to thank in a project such as this. Chief among them is William Kaye Sessions, who asked me to write an update of the research he carried out on York almshouses in 1946.

I am indebted to Linda Haywood of the Bishopthorpe Local History Group, and to John Terry of the Acomb History Society, for providing details of the poorhouses there. Colleagues among the almshouses of York willingly provided information and let me in to take photographs: Alan Scott and Julian Bedford (Burrill's), Andrew Hill and Alfred Fletcher (Colton's), Rosemary Stubbs (Fothergil's), Cathie Halliday (Lady Hewley's), Sheila Hemingway (Walker Barstow), and Edward Waterson (Heslington Almshouses). Everyone at Wandesford House has supported me, particularly Richard Parkin (architect and trustee) who drew the maps for the book. Mandy O'Sullivan at Quacks marked them up with great skill in addition to working on the layout. I am deeply grateful to them all.

Several people offered advice and references, among them Professor Christopher Norton, Alison Sinclair, and also Professor Bill Sheils who critically read and commented on an early version. Any remaining errors are my own. To my husband Lewis I owe thanks not only for reading every version of the text, but also for walking the walls, streets, and alleyways of York with me in search of former almshouse sites and buildings to photograph. He also tolerated my long absences from the domestic scene.

The photographs are mine unless otherwise credited. For the use of three photographs, thanks are due to Joel Kerry of the Archives and Local History Dept of the City Library (http://www.york.gov.uk/archives). Others in the collection may be viewed at Imagine York (http://www.imagineyork.co.uk). York Art Gallery, part of the York Museums Trust, has given permission to use two illustrations, thanks to Jenny Alexander. York Architectural and York Archaeological Society's Evelyn Collection has also given generous permission to use copies of several photographs, thanks to Ian Drake. I have been unable to trace the original copyright owners of some of the Nicholson drawings, but I should like to express my gratitude to them nevertheless.

In the course of my research I used the resources of several libraries in York, chief among them the Minster Library which remains one of the few really quiet library spaces left in which to work. The quietest space of all is the Internet, to which I owe more than virtual thanks for enabling me to access texts otherwise available only hundreds of miles away.

Carole Smith
York

Introduction

The way society has responded to the needs of people suffering hardship, sickness and distress has in some ways changed very little over the last thousand years. This is a survey of how poverty generally has been regarded over this long period. It also shows how the division of the poor into deserving and undeserving affected the relationship between the giver and the receiver of charity. Almshouses were one early response to the needs of the poor on the part of private charity. The state's response was often less benign, and publicly-funded poorhouses and workhouses are also touched upon here.

York's almshouses are the focus of this study and the second half of the book is a gazetteer containing descriptions of the fifty best-recorded of York's almshouses and hospitals. These descriptions include how day-to-day realities of life in them have changed within living memory.

The spur to this study came from William Sessions, who in 1946 at the age of thirty contributed to a national survey of poverty in old age. Carried out by the Nuffield Foundation, and chaired by Benjamin Seebohm Rowntree, it was one of several responses to a postwar concern about the aging of the population. The report of this 1946 survey, entitled *Old people*, was published in 1947. Bill Sessions' two contributions were published as appendices to the report. One looked at York charities (appendix 2); the other reported on the condition of York almshouses (appendix 13, included at the end of this book). His keen interest in the subject continues to this day, and he is the "onlie begetter" of this project, which is an update – or at least a sequel – to his original research on York almshouses and charities. It is an opportunity to bring together a large amount of scattered, mainly secondary, material on almshouses in the city, twelve of which are still successfully functioning.

Charitable giving is closely (though not uniquely) linked to the early history of Christianity, and almshouses share their history with several related institutions established by the early church. One of the actions of the first Council of Nicaea in 325 AD, (Canon LXX), was to send the bishop-delegates away to construct hospitals, another name for almshouses, in every cathedral town in Christendom for the charitable support of the poor, the sick, widows and strangers.

The duty of the wealthy to give to the poor was for centuries backed up by religious belief and biblical exhortation. The rich were seen as stewards of wealth, not owners of it (as in Acts of the Apostles, 4:32) so it had to be shared. Social pressure was applied particularly on anyone who failed in this moral imperative, and clearly it was not entirely effective. At the beginning of the third Christian millennium, it is difficult to conceive of the poverty and misery, the sickness and destitution that existed in European society during the last two millennia. Indeed in the UK, after just over sixty years of a welfare state, the majority of the population can barely conceive of the poverty that existed between the two world wars. It is also difficult in a more sceptical age to appreciate the strength of

Christian exhortations in the past. Nevertheless the charitable urge to help the vulnerable and dispossessed, the poor and the old, has remained. The welfare state, for all its endeavours, has never been able to support all those in need, and charities continue to fill the gap. Giving to charity is still part of many ordinary people's financial arrangements, though it may be rare to find it on a scale comparable to that needed to establish a hospital. In the UK that scale of giving has become a duty of the state and, like general financial aid to the poor, is paid for from general taxation.

It is difficult to rouse compassion for abstractions like "the starving millions", but private charitable donations which follow natural disasters are often enormous. When a desperate need arises and we can see it, we respond. The intimate, personal nature of responses to perceived need is a constant in history. However, sympathy for the able-bodied poor is often less forthcoming. The unemployed poor are supported out of taxation, which is compulsory and annoying. Resentment means that the poor are demonised as feckless scroungers and losers, living a life of luxury at everyone else's expense and laughing as they do so. This attitude to the feckless and the needy alike, without discrimination, continues a pattern of grudging response to need which became established after the Reformation. The feckless, like the poor, are always with us, but even fecklessness is not a key cause of poverty. Antipathy to the able-bodied poor extends in our day to foreign workers attracted by the possibility of earning good money and they are disparaged as economic migrants, as if it were a crime to seek a living. Some are locked up as illegal immigrants, rather as parish outsiders were in the past. This too is part of a pattern: poverty itself was in some periods regarded as a crime to be punished by imprisonment. Worse, refugees tarred with the same brush may be turned away or rounded up and sent back into danger.

In the context of modern charitable giving and state welfare, almshouses are something of an anachronism not least because of the name, which is archaic and much misunderstood. Most people look blank when the word is mentioned, or hear it as "armoury". But almshouses have a very long history, dating back to at least the tenth century in England. Almshouses and hospitals once served a variety of purposes advocated by the medieval church for the benefit of the needy, the old and the sick. Included in these groups were children, widows, wayfarers, the insane, prisoners, and even the dead. The medieval range of welfare provision was broad and apparently generous. After the suppression of the religious institutions in the sixteenth century, welfare provision was necessarily dispersed to private charity, parish overseers and municipal corporations. What followed was the sad story of the poor laws passed by parliament to manage the problem of poverty.

There remain about 2,600 functioning almshouses in the UK, still fulfilling something of their original purpose. Since the sixteenth century they have for the most part been run by trusts and charities; some were taken over by city corporations; others were run by local parishes; some have become housing associations or registered social landlords. They once played an important role in alleviating poverty, particularly in the pre-industrial period. The role they play now is to support around 36,000 elderly people on low incomes to maintain a dignified independence in their own home, supported partly by the state and partly by charitable organisations.

In York a dozen remain and another is in the process of re-establishing itself. Together they accommodate up to 180 people. The oldest almshouse still functioning was founded in Heslington more than four hundred years ago; the newest – inside the city walls – is less than thirty years old.

Tablet recording the refoundation of Sir Thomas Hesketh's Hospital, Heslington

This hospital was
founded by
Sir Thomas Hesketh Knight
Anno Domini 1608.
Removed and rebuilt by
Henry Yarburgh Esquire
Anno Domini 1795
Thou O God hast of thy goodness prepared for the poor

Beneath: "Repaired and modernised by
Richard de Yarburgh-Bateson, 6th Baron Deramore AD 1968"

Burrill Homes, Water End, Clifton

Part I

1
The Medieval Period
Origins of Almshouses and Hospitals

Almshouses and hospitals were originally attached to cathedrals, churches, and monasteries, where alms in the form of food or money were doled out to the poor, sick and destitute – people who often needed accommodation as well as other support. Religious institutions, and also the craft guilds, responded to this need by building refuges specially for people unable to support themselves by reason of age (including extreme youth) or infirmity.

According to tradition, the first purpose-built almshouse in England was established in York in 936 AD through a gift of King Athelstan, grandson of Alfred the Great and the first king of all England, to the canons of the Anglo-Saxon cathedral of St Peter (York Minster). This is one of many myths associated with medieval almshouses, but one with a kernel of truth. The endowment enabled a small almshouse or hospital to be built to the west of the cathedral. Dedicated initially to St Peter, it later moved to a larger site within the south-west corner of the Roman fortress (near the multangular tower) and was renamed St Leonard's Hospital. It became the largest hospital in medieval England. The myth and the hospital are described in more detail in Part II.

St Leonard's Hospital ruins from the Museum Gardens, York

Other candidates for the honour of being the first include the still-functioning Hospital of St Oswald in Worcester. This hospital is said (probably spuriously) to have been founded in the late tenth century by Oswald himself, who was an Archbishop of York. Its origins may well lie in the tenth century, but records only go back to the thirteenth century. The Hospital of St Cross, Winchester, whose financial scandals famously feature in Anthony Trollope's *The warden*, was founded in the 1130s and may be the oldest still-functioning almshouse.

Definitions

Alms is a word now barely recognised by most people. It comes from the Greek *eleēmosunē* - pity, mercy, compassion; a word in biblical exegesis also conveying a sense of sacrificial service to the poor and needy. The English adjective eleemosynary is still occasionally (if rather rarely) used in the context of charitable giving. The Greek word is also the root of words for the concept in other European countries: in Italy *elemosina*, in Germany *Almosen*, in Spain *limosna*, in France *aumône*.

The word almshouse came to be mainly associated with guilds and municipalities, and into common use chiefly to mean refuges for the poor, elderly and infirm. Almshouses were godly houses, whose occupants were expected to pray for the souls of their benefactors. Other names for them therefore include bedehouse, which comes from the Old English word for prayer, *bēde*. Their occupants were called bedesmen, or beadsmen. Other examples of the word are to be found in the York lane, Bedern, where the former chapel of the Vicars Choral survives; in a rosary the beads signify prayers; and even the Venerable Bede may have been so named because he had been intended for the church from birth.

The French term *maison dieu* was often used for smaller establishments; also *hotel dieu*, and occasionally the Latin *domus dei* and *domus elemosinarie*, of which "almshouse" is a direct translation. There is also a rarely used Greek word: *xenodochium* or *xenodocheion* (the Greek for stranger or visitor is *xenos*). In the early church this meant specifically a house for receiving poor and infirm pilgrims, and was mainly associated with monasteries; it was used in this context by Alcuin in a letter to Eanbald II, Archbishop of York at the turn of the ninth century. It was used later in a slightly archaising way to describe an almshouse or a visitors' room in a big house, and Francis Drake used it of St Catherine's Hospital, York, in his history *Eboracum*.

The *maisons dieu* were a chiefly urban phenomenon, often associated with guilds and private benefactions. They were usually very small and unendowed and became more common in the period after 1350 when the middle-classes made bequests for their establishment and continued support. Though there were many of them, they tend to appear only fleetingly in the records of the period – in wills, property deeds, or civic records. (The spelling in these documents often reflects the English pronunciation, for example "masyndewe"). Mostly they were poor and could only offer a bed, or a room with a bed, and perhaps a small stipend to the 'impotent poor' they sought to help, not the total provision of larger foundations; the occupants probably had to beg in the streets. They enabled people of limited means to help the poor, and their continued existence depended on gifts and bequests from other benefactors.

The *Maison Dieu* is, curiously, the name of a tarot card. Also known as The Hospital, and The Tower, it is apparently an unlucky card. This otherwise inexplicable association with the occult may be simply due to a corruption of *deuil*, sorrow, for *dieu*, God. However, it is also possible that the reverse is true and that when Christian hospitals were established the name was changed from *deuil* to *dieu*. There was, for example, a pre-Christian hospital founded in pagan Ireland, around 300 BC, which was called a *Bróin bearg* (or *Brón berg*), house of sorrow. The link with tarot, therefore, remains unexplained.

An oddity among variants on almshouse is "Callis", which is found attached to the names of almshouses or hospitals in wool staple towns: one example is St Peter's Callis in Stamford, Lincolnshire. The association is with Calais, once an "English" town, (and pronounced "Callis" even in Pepys' day) where in the fourteenth century an English company, the Merchants of the Staple, had a monopoly over English wool exports. Its wealthy members founded almshouses, or Callis houses, in wool towns in England.

A related institution was the hospital. These had existed in many cultures for centuries before the Christian era. The word originally meant any kind of refuge, not necessarily one offering medical care though it often provided it. In monasteries the normal place for the sick was the infirmary, where not only care was provided, but cure was attempted. The word hospital comes from the Latin *hospes*, a guest. Its meaning in the medieval period was therefore closer to hospice, hostel or hotel, a place providing shelter and hospitality. Providing hospitality was a significant element among monastic obligations: chapter 53 of the Rule of St Benedict stated that guests should be received as if they were Christ Himself: "I was a stranger and you took me in" (Matthew 25:35). Welcoming the stranger is (or was) common to many cultures. In Greece, which traditionally still welcomes the stranger, it is called *philoxenia,* love of the stranger.

The Normans brought the hospital proper to England – there is a strong association between the establishment of hospitals and the Norman boroughs – and its history is tied to travellers' needs. It is also linked to the military orders of crusader knights – such as the Knights Hospitaller Order of St John. Among the many follies of the crusades was the lost opportunity to meet the Muslim world peaceably and learn how to use Islam's advanced technology and medical knowledge, which would have served Christendom (and its hospitals) more effectively. It was many centuries before the West caught up with either.

Hospitals were often built like chapels, and provided beds down the sides for the old, sick and bedridden, placed in such a way that the occupants could see the chapel altar, and could participate in services. The remarkable *Hotel Dieu* at Beaune in Burgundy shows how this layout worked as, on a smaller scale, does the undercroft of the Merchant

Adventurers' Hall in York. It can still be seen in the design of a traditional (medical) hospital ward, where the nurses' desk takes the place of the altar. Later, divisions were introduced between beds, creating separate cells and providing personal space and privacy. This architectural formula can still be seen as the basis of some types of almshouse design. Many almshouses originated as hospitals which is why they often retain the name.

The contraction "Spital" or "Spittal", which appears in many place names, is an indication of the former presence of a medieval hospital or almshouse. One well-known example is Spitalfields in the east end of London, an area of land formerly attached to St Mary (Ho)spital, founded in 1197 on the east side of Bishopsgate Street. Other examples include several oddly named villages signposted along the A15 south of the Humber which have long puzzled and amused travellers, among them Spital-in-the-Street, and Owmby-by-Spital, whose names derive from the site of a hospital on the old Roman road, Ermine Street, which linked York, Lincoln and London. Such hospitals, or refuges, were often established along major routes to accommodate wayfarers and provide protection from wild animals.

There were other types of accommodation which are sometimes confused with almshouses, such as chantry cottages, whose rents supported a priest or chantry in a nearby church. Examples of these in York are the line of fifteenth-century cottages next to All Saints North Street, and fourteenth-century Lady Row next to Holy Trinity Goodramgate. The (now demolished) former chantry cottages next to St Martin's in Coney Street were used as virtual almshouses and are described in Part II.

Another form of accommodation was the poorhouse, established under the poor laws, which was publicly-funded housing for the poor. Though they fulfilled a similar function they were not almshouses as such: they were not independent or endowed, they did not necessarily provide a pension to carefully selected occupants, though their rent might sometimes be paid from the poor rate, and they were managed by parish overseers of the poor. Two of these just outside York, in Acomb and Bishopthorpe, are described in Part II.

Lastly, "charity" is a word which will recur throughout this book. Biblical translation of St Paul's Greek word *agápē* as "love" in 1 Corinthians is slightly inadequate. Greek distinguishes between spiritual and physical love with *agápē* and *eros* as Latin does with *caritas* and *amor*, but the English word "love" does not. The King James Bible translates *agápē-caritas as* charity. The words are defined by the actions they represent, and charity is a very necessary term for a very necessary activity.

The role of the hospitals

The early role of hospitals and almshouses was to provide refuge and assistance to anyone asking for help at their doors. By the fifteenth century changes in almshouse design suggest that they had become homes for beneficiaries who remained resident in them, rather than hostels for casual arrivals. Selection for a place required certain criteria to be fulfilled.

A poem of around 1536 by Robert Copland, entitled *The hye way to the spytal house*, contains a conversation between the poet and a porter about the inmates of what is assumed to be St Bartholomew's Hospital, London. Copland was mostly interested in describing the type of vagabond who tried to get in, rather than the inmates themselves, but he does list the criteria for selection of people deemed worthy of care. These were, for example, people so ill they could not work, and had no friends to help them; sick and impotent (i.e. frail or powerless) old people; women in labour; people suffering from disease, or injured by violence; honest folk fallen into poverty through no fault of their own; wayfarers and maimed soldiers; and the bedridden unable to beg for alms. If they died, the hospital would bury them properly. But the hospital would not accept healthy, stubborn knaves, or vagabonds, who were regarded from earliest times as undeserving of charity or alms.

Medieval hospitals sometimes specialised among these types of need, and were often dedicated to the saints associated with different sorts of problem, including Saints Bartholomew for skin problems – his martyrdom was, traditionally, being flayed alive; Giles for cripples, the poor, lepers, and nursing mothers; Mary Magdalene for lepers; Leonard for nursing mothers; Anthony, James, Lawrence and Nicholas for travellers, and various diseases. Holy (or Saint) Cross Hospitals may have related to the crusades, and therefore travellers: the Winchester Hospital of St Cross still provides a wayfarer's dole of bread and ale for anyone with the cheek to ask. Some hospitals, a very few, cared for the insane, the most famous being the Hospital of St Mary Bethlehem, or Bedlam, in London. This priory, founded by a military order in 1247, now lies beneath Liverpool Street Station; its hospital was established in 1337, moved several times and now survives as part of the South London and Maudsley NHS Foundation Trust.

Several medieval hospitals practising medicine in London survive as medical hospitals to this day, among them St Bartholomew's (Bart's) Hospital, which dates from the twelfth century. John Mirfield a doctor at St Bartholomew's wrote a medical treatise *Breviarum Bartholomei* in about 1380, using Greek, Latin and possibly Arabic sources. In York too, the sisters at St Leonard's not only offered care to pregnant women and the chronically sick, but they also attempted to cure cases of acute illness.

Some two hundred leper hospitals were established in the Middle Ages for an English population which was, at its greatest, six million but which probably averaged two to three million. They were also known as lazar houses, after St Lazarus, the patron saint of lepers. The Military and Hospitaller Order of St. Lazarus of Jerusalem originated in a leper hospital in the twelfth century and was established to treat victims of the disease, setting up hospitals across Europe. Saint Lazarus was probably the poor man at Dives, the rich man's gate, in Christ's parable of redemption through poverty (Luke 16:19-25). There are accounts in the synoptic gospels of Christ's cleansing of lepers, and the identification of the disease with Christ made it an important object of charity. Leper hospitals were also sometimes dedicated to St Mary Magdalene, perhaps through association with another Lazarus, her dead brother, who was brought back to life by Christ (John 11:1-46).

The association of leprosy with the crusades is likely to be unwarranted – it may have been first brought to England by the Romans. It seems to have been a rampant disease from early times until the fourteenth or fifteenth century when it began to decline. When it did the leper hospitals generally turned to the care of the old and infirm.

Analyses of excavated leper burial grounds indicate that the disease was quite well diagnosed, but some of the people thought to be suffering from leprosy may, of course, have had other diseases. Leprosy is related to the mycobacterial form of tuberculosis, which includes scrofula; there may also have been an unrecognised form of syphilis which would have presented similar symptoms. Skin diseases generally were common in the medieval period, owing to proximity to animals and vermin, lack of clean clothes, and low resistance to infection. One of the most important actions of the hospitals was to provide clean conditions. Clothes and bedding were washed, even boiled or baked if verminous.

Aging and maimed soldiers, another needy group, would have been taken in by any hospital, like other aged or infirm people. Specialist hospitals caring for old soldiers and sailors, and their dependants, tend to be a later phenomenon, Chelsea Hospital for example dates from the seventeenth century.

The care of children was a special consideration; it was also important because of Christ's welcoming of them. Hospitals attached to religious foundations sometimes took in foundlings or looked after orphans born in them. Their functions therefore increasingly turned to education and learning. Many monastic and hospital foundations had schools of theology, and sometimes of grammar, which continued even after the establishment of the universities of Oxford and Cambridge. They educated the children, and in some places also trained them as choristers. After the Reformation some of those hospitals became grammar schools (such as Christ's Hospital in West Sussex), and two became Oxbridge colleges (St John's College, Cambridge, and Magdalen College, Oxford.)

Few medieval hospitals were as large as St Leonard's in York or St Bartholomew's in London; most took in around twelve or thirteen – a biblically significant number, and one that was also manageable. The religious brethren and sisters caring for the inmates were most likely to be members of the Augustinian rather than the Benedictine Order, because the Augustinians required less time to be devoted to prayer. St Leonard's belonged to the order, and was close to the Augustinian Friary which had the neighbouring river front. The related Gilbertine order also engaged in this kind of work.

St Leonard's Hospital ruins from inside the Roman fortress, York
Undercroft of the infirmary or chapel

St Leonard's Hospital ruins from inside the Roman fortress, York

2
Medieval Charity

Before the Reformation, a Christian's relationship with Christ was governed by the injunction to love one's neighbour, and acts of charity were defined by canonical rather than secular law. An act of Corporal Mercy which improved the lot of anyone less fortunate than oneself was charitable and pious, of benefit to one's soul. The virtues of charity and piety were inseparably linked. The poor, being guaranteed entry to heaven unlike the rich, could reciprocate by pleading for benefactors on the grounds of their charitable giving. There was also a concept of the stewardship of wealth, by which the rich understood that they were not absolute owners of wealth, only God was. They therefore had to share it with the poor and needy.

Poverty itself was not regarded as a crime, and destitution was seen not as a form of moral turpitude but more like an illness coming out of the blue. The poor were therefore regarded with compassion, not with the suspicion and moral condemnation that prevailed after the Reformation. This may also have been because the medieval population was small and composed of small communities in which there were strong social structures and where social controls were easy to maintain. Compassion for those falling by the wayside arose from sometimes personal knowledge of the individual, but also from a sense of humility before God. However, there was resistance to supporting those seen as undeserving. A distinction between the deserving and undeserving has always been made by benefactors, even if some of the poor were unfairly judged.

The Seven Works of Corporal Mercy

The Seven Works of Corporal Mercy come from biblical texts, the first six from Matthew's Gospel, 25:35-36: feeding the hungry, giving drink to the thirsty, clothing the naked, receiving the stranger, visiting the sick, and comforting the prisoner; the seventh, burial of the dead, comes from the Book of Tobit (1:16-17) in the Apocrypha, and was apparently of less interest to most benefactors. There is a stained-glass window in All Saints North Street, York, containing a depiction of the first six Corporal Mercies, including a (rather unconvincingly) benign character giving money to a group of men in the stocks.

The Doomsday play from the York cycle of mystery plays makes it plain that Christ was to be found among the poor, and that salvation or damnation depended on an individual's fulfilment of these works of mercy. Pat Cullum (1989) cites the interesting case of a chaplain in York, Richard Kirketon, who had not only endowed a bed in John Holme's *maison dieu* in Whitefriars Lane, but also erected (or more likely dug) latrines in a garden, possibly for the use of the poor – an act of extra-ordinary corporal mercy which was no doubt welcomed by its beneficiaries. In his will of 1486 he left the garden for the use of that *maison dieu*, and another for the poor in John Bedford's *maison dieu* in St Andrewgate.

Many gave money in their wills to be distributed to the most needy – sometimes specified by type of need, sometimes not. The giving out of doles where no limitation had been placed on types of recipient frequently attracted enormous crowds. However, there was often a personal relationship between donor and recipient which was important to both: probate records show that many bequests were for sums to be paid regularly to named individuals – who might be inhabitants of a hospital, or someone regularly helped while the benefactor was alive.

Comforting the prisoner, from the Corporal Mercies window
All Saints North Street, York

The blind, not often taken into hospitals, were frequently remembered in this way, as were the lame, the bedridden, poor widows and prisoners.

It was part of a housewife's responsibility to give alms. Day-to-day charitable giving by ordinary people often involved the women of the house receiving beggars or wayfarers at the door and giving out regular doles of food. They might even take half a dozen of them in for a meal once a week. Some undertook to house such people overnight on a regular basis – something that might have ultimately led to the establishment of a *maison dieu* as a way of providing several works of corporal mercy in one go.

Merchants sometimes bequeathed quantities of their stock-in-trade for distribution to the poor. Where food and drink were concerned, such wills might leave quantities of grain to be made into bread or ale for the poor. Some left food for a funeral feast, or for a regular feast at a hard time of the year.

Clothing was frequently distributed at funerals; and garments, or materials from which to make them, often featured in bequests. They were usually of russet, a coarse homespun fabric, reddish brown, grey or neutral in colour. Sometimes they were black, to symbolise mourning, when the recipients needed to be appropriately dressed to attend the body at the mass or to the grave; and sometimes white, to emphasise penitence, particularly for almspeople. Sumptuary laws governed the quality of the materials.

Specific bequests were made for the general upkeep of the inmates of hospitals, *maisons dieu*, and prisons. The injunction to visit the sick seems to have been largely replaced by gifts to the hospitals housing them, particularly the leper hospitals. Prisoners, too, were remembered and often relied on charitable gifts to survive – it was not unknown for people to starve to death in prison. St Leonard's Hospital provided bread to the prison in York, presumably to prevent this kind of thing and because it did so there were fewer private benefactions for prisoner relief in York.

Burial of the dead attracted few specific benefactions, but the confraternities or guilds sometimes arranged burials, so it may be that donations they received were used for the purpose. Instead, other kinds of doles were considered, often of fuel. Women, especially, made this kind of provision, both to *maisons dieu* and as outdoor relief. Some benefactors' wills helped with the remittance of debts (particularly debts to themselves) on the part of individuals and hospitals. There were bequests of dowries for poor girls to prevent them falling into prostitution; and of money to support, among others, poor or dependent relatives, aged priests, poor widows and poor scholars

The repair of roads and bridges was a frequent and useful object of benefaction, and fulfilled the charitable purpose of helping one's neighbour and providing a service to many.

Financial maintenance of hospitals

Bequests to charity remained an important duty, and wills of the period demonstrate that charitable bequests sometimes accounted for between a third and a half of the value of a testator's property, and where there were no heirs, it might be even more. Property was often left in this way for the support of a hospital. Some may have thought that to give away a large proportion of their wealth at death might also allow them to become poor at the right moment, enabling them to enter heaven. But lifetime giving was regarded as of more likely benefit for the soul. Many hospitals, almshouses and *maisons dieu* received this regular form of support, enabling them to feed and clothe their destitute inmates and pay them a stipend. Beds could be endowed for use by the poor, or for future use by the benefactor. If for the poor, it became one of the Seven Works of Corporal Mercy bringing benefit to the soul, and a poor person appointed to an endowed bed would be expected to pray for the soul of the benefactor. This form of support was cheaper than endowing a whole hospital, and benefactors making bequests to hospitals could claim various rights including that of burial in the hospital. However, there was occasional conflict, particularly where benefactions produced a right of accommodation to the giver, because there was a temptation to abuse it. Some hospitals complained that patrons used them as free hotels for themselves and their retinue.

The large hospitals, like the large monastic foundations, needed substantial and dependable resources for maintenance. They might receive royal and papal grants; they also often owned enormous amounts of land with farms and granges from which they could derive not just food but also rents: St Leonard's, for example, had sheep, pig and dairy farms as well as mills.

Hospitals, and church-owned property generally, were not subject to taxation or payment of tithes, so the King took an interest in how his subjects left their money in this respect. Magna Carta included some legislative restraint, and at the end of the thirteenth century two statutes of mortmain were enacted, which prevented land from passing into the hands of the church and thereby out of the feudal economy. Mortmain was a prohibition against perpetual ownership, and is sometimes defined as the dead hand that leaves its wealth to a body that never dies – that is, the church; so by ensuring that valuable estates were not alienated at the expense of the exchequer, the statutes of mortmain preserved the kingdom's revenues. Loopholes in the law were continually subject to patching legislation thereafter, including the Statute of Uses 1535, the Statute of Wills 1540, and the Statute of Charitable Uses 1601. Many more followed until the establishment of the Charity Commission in the nineteenth century (Chapter 6 below).

Begging

Begging for alms in the streets was a necessity for the poor, but also a crime, so it had to be licensed not to fall foul of the law. The king might grant such a licence to a hospital; an archbishop might do so for people suffering hardship from some sort of accident like losing one's goods through fire, shipwreck, robbery, or sickness. Leper boxes or alms-boxes were sometimes carried about or were hung at the gates of hospitals for casual donations.

People expected to support the poor without discriminating among them. Anyone attempting to get undeserved alms might be treated with disdain but this did not undermine the sense of responsibility felt by those who had much for those who had nothing. And even the undeserving might be fed if they were hungry.

Travellers, pilgrims and wayfarers were free to call at private houses on their way to ask for sustenance. This was not regarded as illegal begging, and bread and ale would be given as a matter of course – these were Corporal Mercies – but the attraction of pilgrimage faded and, at the end of the fifteenth century, social problems associated with the end of the Hundred Years' War began to take effect. There were more vagrants and roving bands of "masterless men" trying to take advantage of the privileges of pilgrims. Genuine wayfarers and travellers may have had trouble proving their honesty, and the type of charity provided at the door for them may not have survived the fear generated by marauding bands of discharged soldiers and other criminally inclined vagrants. Casual arrivals at the gates of a hospital tended to be allowed only one night's lodging and only the chronically sick and women in labour were allowed more than temporary accommodation. A vagrant gypsy life was increasingly discouraged and such wanderers began to be regarded with suspicion rather than compassion. They attracted a good deal of opprobrium and proclamations were often made against them.

St Leonard's Hospital ruins from Library Square

3
Medieval York: Almshouses and Hospitals 936 - 1558

This study is chiefly about York, for many centuries England's second city after London. The Christian duty of almsgiving would have been generally carried out at the gates of monastic establishments in the city and at the cathedral of St Peter. Talk of establishing a hospital or almshouse in York started quite early, in about the eighth or ninth century – Alcuin referred to it in a letter to the Archbishop – but nothing obvious happened until the tenth century when an almshouse dedicated to St Peter (later St Leonard's Hospital) was founded. Detailed histories of St Leonard's and of York's other hospitals and almshouses may be found in Part II.

Poverty in York

William the Conqueror's vengeful, almost psychopathic, "Harrying (or harrowing) of the North" in 1069 was a devastating attack on the population of northern England. William adopted a scorched earth policy and the famine and poverty that ensued brought starving refugees from the surrounding areas flooding into York for help and shelter. St Leonard's would have been one source of aid, and as a result no doubt attracted subsequent donations for its work.

The devastation had social, cultural and economic effects which lasted for decades if not centuries. In common with other towns and cities, York saw much poverty and destitution and became a fairly squalid place, both by modern standards and apparently even by medieval ones. For Edward III, a frequent visitor in the fourteenth century, it was an egregiously smelly city, chiefly because of the keeping of cattle and pigs in the narrow streets. The Victoria County History (VCH 1961) includes gruesome descriptions of conditions in the streets. There were high levels of prostitution too, historically a sure source of income for destitute women. It also gave rise to several street names and landmarks among the slums of the walled city. There were many Love Lanes, for example. One of these led up to a latrine on the city walls, near to Harlot Hill Tower, from the junction of St Andrewgate and Aldwark. It ran between the Merchant Taylors' guild hall and its almshouse, making it a less than delightful spot for the elderly to live in. That particular Love Lane has gone, but the latrine, now fossilised behind a railing and harbouring only cigarette butts, and Harlot Hill Tower are still there as a reminder of past wretchedness.

St Helen's churchyard had a "Cuckold's Corner", which would originally have been overlooked by the Guildhall chapel and *maison dieu*. This spot now lies under the centre of St Helen's Square overlooked by the Mansion House. The only remaining queues in the area are for Betty's Tearooms. Grape Lane, a less graphic version of the original name, and the streets and alleys around it, such as Mucky Peg Lane, would also have been an area to avoid if possible. Whether the prostitutes qualified for charitable aid is open to doubt, though some hospitals took in poor women who were about to give birth, and cared for their orphans if the mothers died. One *maison dieu* near the 'red light' district off St Andrewgate, where the population was predominantly female, took in only women throughout its existence.

The city walls at the end of Love Lane
(latrine)

Hospitals founded in York

St Leonard's, almost certainly the first and at one time probably the largest hospital in England, housed and tended the sick of all types and ages, particularly pregnant women, and looked after the orphans of destitute women dying in childbirth.

There were also four, possibly five, leper hospitals on the roads leading into York, which by 1400 is thought to have had a population of around 15,000. St Nicholas's Hospital, the largest and with the richest endowments, was founded in the 11th or 12th century in Walmgate Without, now Lawrence Street. Another, smaller, leper hospital dedicated to St Leonard (sometimes cited as St Loy) was founded outside Monk Bar, beside Monk Bridge, in probably the 12th century. St Catherine's was founded in 1333 outside Micklegate Bar in what is now Blossom Street. St Helen's "spittle house" was known to be in existence in 1399, outside the city walls in Fishergate. The Hospital of St Mary Magdalene at the city boundary in Clifton, near the Burton Stone, may well have been a leper hospital, given its position and its dedication to the sister of Lazarus. It is also thought that lepers were accepted in other houses, for example the civic *maison dieu* on Ouse Bridge.

This was a period of pilgrimage to holy shrines, such as that of St Thomas Becket at Canterbury. Hospitals for poor wayfarers and pilgrims provided much-needed, and often free, accommodation on the way. Chaucer's Canterbury pilgrims were well off and, rather than receiving free accommodation at monasteries and wayfarers' hostels, were more likely to have paid their way. York Minster had a shrine to its own saint, Archbishop William FitzHerbert, and the most famous hospital for wayfarers in York, also a hospital for the poor of the area, was St Thomas the Martyr's just outside Micklegate Bar.

The number of hospitals and almshouses in York in the medieval period was put at twenty-two, and in London about twenty-five, by Rotha Clay in 1909. These figures are probably an underestimate for each city. It is likely that there were between thirty and forty in York and perhaps also in the city of London, though the number of hospitals around London specifically for the sick may have been very much larger. William Aycliffe (2008) suggests that sixty-eight were founded in a ring outside the city of London between 1070 and 1150, half of them for lepers.

Guild *maisons dieu* and hospitals

York has four surviving medieval guild halls, three of which still have their associated almshouse, *maison dieu,* or hospital buildings, though these have long-since ceased to function. The four are the Merchant Adventurers, the Merchant Taylors' and St Anthony's Hall, Peasholme Green; the fourth, the Common Hall (now the Guildhall), gutted during World War II, has since been restored for continued use by the City Council. The guild of St Christopher moved to the Common Hall from Fishergate Within in 1456, and its *maison dieu* was possibly housed in or under the chapel. The Mansion House now occupies the site.

Other guilds, too, made provision for their elderly members. The guild hall of the Cordwainers Company (of shoemakers and leatherworkers) was in Hungate; later turned into tenements for the poor, it was demolished in the nineteenth century. Its *maison dieu,* founded by John Marton in 1436, is generally located by historians to the area of Walmgate Bar and St Denys' church, in what was or became Church Lane, off Walmgate. Rebuilt, possibly on the same site, in 1811 it survived into the twentieth century but was demolished in about 1938. The Butchers' Guild may have had a *maison dieu* in Little Shambles, perhaps in the building it used for a guild hall. There is also mention in the records of a Tanners' *maison dieu* in the Tanner Row/North Street area, but it is not clear whether it was actually built or established. Guild-maintained hospitals and *maisons dieu* generally survived the suppression of the religious houses in the sixteenth century.

St Anthony's Hall, Peasholme Green

Private *maisons dieu*

These usually small houses, often founded or taken over by private benefactors for the poor, feature only fleetingly in the records, and sometimes confusingly. Where they are mentioned in wills, it is often merely as, for example, "to Katherine ... in the maison dieu". Sometimes the original founder's name appears in the bequest, and so, occasionally, does the street name, which may have changed out of all recognition even if the street remains in existence. The people taken into *maisons dieu* seem to have been in particular the elderly, but also poor travellers, immigrants or pilgrims.

Rotha Clay (1909) lists several *maisons dieu* founded by private individuals, and locates them variously in Peter Lane, Hestergate/ Hertergate, Markyate, St Andrew's Lane, North Street, Ousebridge, Whitefriars Lane, Layerthorpe, Monk Bridge, Micklegate, and Fishergate. T.M. Fallow (VCH 1913) lists others on Layerthorpe Bridge, Little Shambles and St Andrewgate. Elizabeth Brunskill (1960) muddies the field further with a few others which may be partly due to a confusion of the names of locations. She also refers to one *maison dieu* which seems to have been either founded or remained in existence in York after the Dissolution in the sixteenth century. It was in Fetter Lane, and is mentioned in the will of Robert Hogg in 1558. Pat Cullum (1989) suggests there were between twelve and eighteen *maisons dieu*, mostly founded around the turn of the fourteenth and fifteenth centuries, during an economic boom after the Black Death, when more people were in a position to help those less fortunate than themselves. Her analysis of the probate records provides a better source for the locations of these small establishments.

The duplication of locations in various lists seems to suggest that, for example, there were two or more *maisons dieu* in the St Andrewgate area, two near Monk Bridge and Layerthorpe, and several in the Fishergate area (inside the walls). There is so little known about them from the records that positive identification cannot always be made, but the large number contained in these lists suggests that there was an extraordinary level of need.

Maisons dieu founded by private individuals were often passed to the mayor and corporation of the city to be managed after the death of the founder. The founders often had links with the corporation, so this would have been a natural course to take as it also ensured their survival. The one on Ouse Bridge had close links with the corporation, and had probably been intended as a civic hospital.

Though most of the houses were very small, those founded by the wealthy cloth merchant Robert Holme, his brother Thomas, and John de Craven seem to have accommodated between twelve and as many as twenty. Many of those listed as founded by men were commonly referred to in wills under the names of their wives, which has led to the latter being listed separately, even under a slightly different spelling of the surname. This also compounds the confusion about how many there were.

Some of the streets and lanes where they were located have disappeared. Whitefriars Lane, named for the Carmelite Friary nearby, may have been near or on the line of what is now Black Horse Passage. It ran off Stonebow Lane which linked Fossgate to Hungate before the construction of the modern Stonebow. It may alternatively have been an old name for Stonebow Lane. Peter Lane, or Peter Lane Little, is off High Ousegate; the name refers to the lesser church (as opposed to cathedral church) of St Peter that once stood in the area. The parish was united with All Saints Pavement, and the church was suppressed in 1586. The location of Markyate is possibly Monkgate, according to Hargrove (1818). Hester- or Hertergate, was later known as Far Water Lane, and part of it survives as modern Friargate. It was a lane running down the side of the Franciscan Friary from Castlegate to the river, later a wretched, disease- and crime-ridden area, though highly picturesque – a watercolour of it appears in Buttery (1984). The construction of Clifford Street in about 1878 cut across the Water Lanes and altered the entire area. Many of the timber-framed houses, squalid and verminous, were demolished and their inhabitants turned out. Bill Sessions' father recalled being shown the site as a child in the 1880s.

The approximate locations of these small establishments are marked with an **[x]** on Map 1: Medieval York (p.69).

Economic decline in York

By the end of the fifteenth century York was in financial decline, and the income of the hospitals was much decreased. There was a falling population, and some neighbourhoods were abandoned which led to a loss of rental income. In addition the rise of the cloth industry in the West Riding began to affect the city's cloth trade, and there was an increase in unemployment. The corporation began to be concerned about the numbers of vagrants coming into the city and settling. A statute on vagabondage enacted in 1495 provided a useful excuse to exert control over an increasing problem. Vagrants were discouraged from coming, by imprisonment initially, but no particular animosity was felt until the problem of dealing with large numbers of unemployed, and probably desperate, people began to affect city life. Many people came into the city from the countryside seeking help and in the early years of the sixteenth century the city took a census of the poor and issued licences to beg. Those with licences had to wear a badge, or be sent packing. Some were sent to an early form of workhouse to weave and spin for their living; those that "would not" were imprisoned. It is not clear what defined a "would not" response.

Tablet recording the refoundation of the Merchant Taylors' maison dieu, Aldwark.

"This Hospital built
by the Merchant Taylors
Company, for the use of
four poor brothers or
sisters, in the year 1730.
Mr Thomas Spooner, Master, Mr John Napier, Mr John Etherington,
Thomas Ellis, Sen., John Dunning, Wardens"

York almshouses and hospitals 936-1558

The following list is compiled from several lists and references, among them: Rotha Clay's list in *The mediaeval hospitals of England,* 1909, T.M. Fallow's in *The Victoria history of the county of York*, vol. 3 (VCH 1913), and Elizabeth Brunskill's in *Some York almshouses,* 1960. P.H. Cullum's doctoral thesis (1989) provides the most accurate researched analysis, particularly of the *maisons dieu*. The cited sources are often probate records, and it is clear that there is uncertainty about much of the detail. Most of these almshouses and hospitals have disappeared without trace.

The numbers refer to locations on maps 1 and 3 (p.69 & 71), and descriptions in Part II

(1) St Leonard's Hospital c936. Originally St Peter's. SW corner of the Roman fortress.
(2) St Nicholas's Hospital 12th century. Lawrence Street. For lepers.
(3) St Giles without Walmgate 1274. Lawrence Street.
(4) St Loy (St Eligius) / **St Leonard** Monk Bridge, 12th century(?). For lepers.
(5) St Mary in the Horsefair 1315/18. The Horsefair (Union Terrace/ Clarence Street)
(6) St Catherine's Hospital 1333. The Mount. For lepers.
(7) Holy Trinity Hospital c1365, licensed 1372. Fossgate: the undercroft of the Merchant Adventurers' Hall.
(8) St Thomas's Hospital 1390. Next to Micklegate Bar. For wayfarers and the poor.
(9) St Helen's fl. 1399. Fishergate. For lepers.
(10) Confraternity of St John the Baptist (Merchant Taylors) *Maison dieu* c1415. Aldwark.
(11) St Anthony's Hospital, Gillygate founded before 1420. The Horsefair (Junction of Wigginton Road and Haxby Road).
(12) The Cordwainers' *Maison dieu* 1436. St Denys's Church Lane, off Walmgate.
(13) St Anthony's Hospital, Peasholme Green, 1446.
(14) Common Hall, St Christopher's Guild *Maison dieu* 1456. Coney Street.
(15) St Mary Magdalene before 1481. Clifton, at the city boundary near the Burton Stone. Probably for lepers; later for clergy.

Maisons dieu

In order of probable date of foundation. Approximate locations marked **[x]** on maps 1 and 3 (p.69 & 71).

Ouse Bridge 1302: re-founded as a city *maison dieu*, originally a guild foundation. SW end of the bridge near or opposite St William's chapel, possibly above the prison on the bridge. For the poor and sick, possibly including lepers.

Little St Andrewgate (now part of Spen Lane) 1353: founded by Thomas (aka Richard) de Duffield for the poor.

Monkgate or Monk Bridge 1353: built by Robert Howm or Holme, brother of Thomas, for the poor.

Markyate (possibly another name for Monkgate) 1353.

Whitefriars Lane and/or Stonebow Lane 1362, 1442, 1471. Possibly two (or three) establishments: one a guild foundation, another founded by John Holme, for the poor and sick, and migrants.

Peter Lane Little 1390: founded by John de Derthyngton/Darthyngton/ Dernyngton for the poor.

Hestergate or Hertergate (now Friargate) on the corner with Castlegate 1390: founded by Thomas Howm or Holme, brother of Robert, for the poor, with an associated chantry chapel where they were prayed for.

Fishergate Within 1391: founded by Nicholas de Skelton for the poor; may have been taken over by St Christopher's Guild.

St Andrewgate 14thC: foundation by Cecilia Plater for poor women.

St Andrew's Lane 1397: founded by Richard Platter (possibly the husband of Cecilia Plater, and therefore possibly the same establishment).

North Street, next to All Saints church 1397: founded by John and Ysolda de Acastre for the poor. May have been run by the Weavers' Guild by 1495.

Layerthorpe Bridge, or possibly on an upper floor of the postern gate, c1407; founded by John de Craven for the poor; passed to the Bigod family who lived nearby. Depictions of the postern appear in Buttery (1984), and Murray (1990).

North Street, down an alley, 1408: established by William de Sallay, for the poor.

Fishergate Within c1436-56: dedicated to St Christopher; transferred to the Guildhall mid-fifteenth century.

Marygate 1436: a frerehouse (probably for elderly monks of St Mary's Abbey).

Aldwark 1439: (Holy Trinity)

North Street or Tanner Row 1446: the Tanners' *maison dieu* if the house actually came into existence.

Fetter Lane, possibly established as early as 1448 (appears in Robert Hogg's will of 1558).

Little Shambles 1470: mentioned in a bequest by Elizabeth Dugdale. Possibly the hospital of the Butchers' Guild.

Little St Andrewgate (now part of Spen Lane) 1485: founded by John Bedford (possible take-over of Duffield's of 1353). For the poor.

Micklegate 15thC: founded by Sir Richard York.

Fishergate [N.d.]: possibly dedicated to St Catherine.

Horsefair [N.d.]: possible hospital, or chantry chapel dedicated to St Anne.

4
Charity in the Seventeenth and Eighteenth Centuries

The aftermath of the Hundred Years' War produced social changes at the end of the medieval period. The rich were taxed to pay for the wars; there was the humiliating loss of French territory; and there were, increasingly, roaming bands of masterless men, some discharged from military service, who terrorised citizens and travellers. It may be no coincidence that the picaresque novel – stories of vagabond heroes, such as *The unfortunate traveller* by Thomas Nashe – began to appear at this time as well as, perhaps more significantly, stories about Robin Hood. These emerged from an earlier ballad form in the early fifteenth century and became popular in the sixteenth century, in a period when patterns of charitable giving on the part of the rich were changing.

After the Reformation, the poor began to be regarded with less compassion. There was an increasingly hard-line attitude taken towards those applying for charity or parish relief who were not the impotent poor, elderly or frail, widows or infants. The poor, particularly unemployed men of working age, began to be seen as to blame for their own poverty and the first pejorative terminology began to be applied to them. Such terms as "sturdy beggars" and "mighty of body and not sick or impotent" included the suggestion that they were idle and workshy, and should be punished – despite the lack of work available. The deserving poor, those unable to work and needing to beg, had to wear a token or badge on their outer garment to prove the genuineness of their claim for alms. The scale of demand put an embarrassing pressure on charitable donors – no-one can fill a bottomless pit of need – and may have led to a demonisation of the unemployed poor. It is easier to blame people for their own difficulties than to admit guiltily that you cannot help. Cullum (1989) even suggests that a similar demonisation of needy old women, characterising them as witches, may also have come from the tensions created by this helplessness and sense of guilt. In York, the reappearance of plague led to its association with poverty and further deprecation of the poor. Gradually the poor, deserving or otherwise, began to experience a more punitive climate. Beggars from outside the city were particularly harshly treated. It was for people like these that the poorhouse and later the workhouse were created by a society unsympathetic to the reasons for such poverty. Those without (or unable to find) work came to be automatically regarded as being the cause of their own misfortunes. This attitude towards the beneficiaries of poor-law aid and charity has lasted until the present day, though it began to be challenged in the twentieth century. The poor were increasingly weighed in the balance, and helped, penalised or stigmatised according to whether or not they had been found wanting.

Charity law and poor law

With the dissolution of the monasteries in the sixteenth century, many associated institutions closed or were in danger of closure – including almshouses, hospitals and schools. The previously varied welfare services provided by hospitals and almshouses in the medieval period probably settled from this point into their constituent parts, with the elderly and infirm being the chief residual beneficiaries living in almshouses. The word hospital is still attached to some of them for historical reasons, but the hospitals providing medical care became very different institutions from this period on.

New legislation under Elizabeth I made it easier to set up endowed charities, which were badly needed to fill the gap left by the suppression of the pre-Reformation establishments: for example, the Act for Erecting Hospitals, or Abiding and Working Houses for the Poor 1597, and the Statute of Charitable Uses 1601. Medieval poor laws were codified. The Act for the Relief of the Poor was passed in 1597 and updated to include provision for the deserving poor in the Elizabethan Poor Law of 1601. Charitable care of the impotent poor became a parochial responsibility. Parishes were required to provide basic food, shelter, and clothing for the legitimately needy living in the parish. The cost was to be met by taxation of wealthier parishioners, which naturally reduced the intimacy of the relationship between giver and recipient. The removal of freedom of choice about giving may also have increased resentment of the poor, and when the latter no longer needed (indeed it was forbidden) to reciprocate with prayers for the souls of benefactors the relationship was further unbalanced.

Other statutes were enacted to prevent abuses and to channel charitable giving to where it was most needed. The Charitable Uses Act 1601, sometimes known as the Elizabethan Statute, was one of many pieces of legislation aimed at policing charities and ensuring that the public benefit they purported to supply was equal to the defrayed taxes; it held to account those entrusted with distributing charitable gifts. The Act also defined and codified what was meant by charitable purposes and who were regarded as genuinely needy.

After 1662, the Poor Relief Act, also known as the Settlement Act, and strictly speaking a "settlement and removal" act (13&14 Car. II, c12), made it necessary to establish everyone's parish of residence, so that it was clear who was responsible for the relief of someone who fell into need. Parishes kept pension lists of those entitled to receive relief. If you left your parish you needed to show your new parish a settlement certificate which guaranteed that the home parish would pay for your removal back there if and when it became necessary. Seasonal harvest workers were exempt (as, of course, were the wealthy). Naturally parishes were reluctant to supply these certificates. This in turn meant that people could not easily move about to find work as, in the absence of such a certificate, other parishes would not be willing to allow outsiders to work there who might become a burden. The Act was repealed in 1834 chiefly because it impeded the mobility of workers. In the 1690s a succession of bad harvests led to widespread distress and there was a series of enactments designed to cut the resulting cost of supporting the poor. Pension lists were regularly revised to reduce the numbers of people likely to demand help. It was probably impossible for the overseers to match resources to need when so many were poor as a result of unemployment, under-employment, short-time working, and had large families to feed.

Private almsgiving and the deserving poor

However, unlike public charity, private charity suffereth long and is often kinder. Many charities were established in this period to help the poor. It was particularly true in York which had a high number of charities from very early on.

Though almsgiving remained a Christian duty, and failure to do so was still frowned upon, it was still the case that the wealthy would not give to people regarded by society as the undeserving poor. As already described, this group included men of working age without paid work who might be living as vagrants, sometimes with their family. Such people were regarded as feckless sturdy beggars. Alms were still generally given only to certain groups in society – essentially people of good character, the deserving poor fallen on hard times and unable to pay their way. Apart from the elderly, sick, and frail, this category included widows, aging spinsters, craftsmen or their widows, sometimes other groups, such as religious dissenters, but also old soldiers or sailors. The Royal Hospital Chelsea was founded in 1682 by Charles II for old soldiers; sailors had hostels or missions to go to in many ports. There was also help for the orphans of, for example, drowned fishermen. Since then other support charities have been established for the armed services and merchant navy.

For some poor children there were charity schools, Blue Coat and Grey Coat schools, which started in London and spread. A great many were established in the seventeenth and eighteenth centuries and aimed to help the children into a trade. Christ's Hospital, now in Horsham, became one. York's own Blue Coat School for boys was housed in St Anthony's Hall, Peasholme Green. The Grey Coat School for girls was originally at 60 Marygate, and moved in 1784 to 33 Monkgate. They were often run and supported by parishes, or by wealthy merchants, and provided clothes and food, and sometimes accommodation. Old charitable tenets die hard: a statuette of an eighteenth-century boy pupil near the church of St Mary Kensington bears the inscription "I was naked and ye clothed me". The blue dye of the boys' coats was the cheapest dye – often used for servants and the lower orders (and almspeople) – and it is said that the saffron used for the children's stockings prevented rats from nibbling their ankles.

But even children were seemingly divided into the deserving and undeserving poor, and the distinction could be harsh on the latter. In the early years of the seventeenth century, some of London's orphans and destitute street children paid the penalty for poverty by being sent to Virginia as plantation workers. One hundred children were rounded up and sent off in 1619.

Patronising attitudes and the effects of more-or-less benign tyranny are demeaning and demoralising at any time, and conditions in some almshouses may not have been so very different from the poorhouse or the later workhouse. Some introduced work of various kinds, like weaving, to enable (or force) the inmates to make a living. However, many also provided a stipend, as well as charity in kind, such as fuel and food, and perhaps a new dress from time to time, some sort of uniform or a badge. This may have been true in York, though only Ingram's Hospital has records showing such provision. The hospital provided fabric for a dress every three years, and a silver badge. This was not uncommon in almshouses elsewhere in the seventeenth century and where there was a special garb almshouse-residents often still wear a version of it for festive occasions. Dress was a general indication of status and occupation, codified in sumptuary laws until well into the seventeenth century. It could be also used to identify groups regarded with disfavour by society – and was used for that purpose even in the twentieth century (for example in Nazi Germany). Whether the dress of almspeople originally also stigmatised those obliged to wear it is not clear, but the Ingram's badge was a little like labelling the deserving poor. Indeed at the end of the seventeenth century there was provision in the settlement laws for the poor to be forced to wear badges to indicate their status, as in previous centuries.

Ingram's Hospital, Bootham.
Doorway from Holy Trinity Priory, c1190

Eighteenth-century poor relief and social control

Thomas Gilbert MP, a poor-law reformer, was instrumental in getting various Acts passed between 1765 and 1782, in particular the Relief of the Poor Act of 1782, which allowed the grouping of parishes into unions whose workhouses would be exclusively for children, the aged and the infirm.

Pauper children were sometimes trained in spinning and weaving, and in the workhouses woollen goods were manufactured from the yarn. Abandoned children were at the mercy of a disapproving society, foundling children being for the most part illegitimate. Christ's Hospital founded in London in 1552, took in foundlings as well as legitimate orphans, but after 1676 the illegitimate were prohibited. The problem became severe in London where one significant piece of combined philanthropy in the eighteenth century produced the Foundling Hospital. Thomas Coram was joined by Handel and Hogarth in this unusually benevolent project – though it required the support of the King and Queen to get it off the ground. The hospital's subsequent history was unfortunately marred by cruelty and corruption. Foundlings elsewhere were usually thrown on the parish where they were unlikely to thrive. In the middle of the eighteenth century, at least two thirds of children (of all classes) in London died before their fifth birthday, but the figure for workhouse children was much higher – possibly as high as ninety percent.

There was, in theory at least, a philosophical urge to philanthropy – good works might do more for your self esteem than good church attendance. But the sheer waste of life apparent in the slums and poorhouses of London also had an impact on the naturally benevolent. Private philanthropy was separate from public poor relief and state provision generally, and has remained separate. It has always been significant in the alleviation of poverty, but has often been regarded with suspicion by the state and subject to legislation – sometimes justifiably, there was certainly corruption. Nevertheless, in the following century private benefactors took on some extraordinary charitable works.

The social controls established by the poor laws at the end of the sixteenth century, with their criteria for how the poor should be relieved, continued through the eighteenth and into the nineteenth century. There were some like William Cobbett, who defended the interests of the poor (and was as a result regarded with grave suspicion), but there were others like Thomas Malthus whose *Essay on population* suggested another kind of control, the elimination of the problem, rather than encouraging it by providing charity. Evangelical Christians too, though claiming to support the poor, thought almsgiving merely encouraged idleness and profligacy. Instead they wanted to exercise more social control by intruding on the poor, keeping them in their place and generally preventing them from behaving in what they considered to be (and probably were) morally dubious ways. The idea that giving to the poor produces apathy and encourages idleness and dependence is one that runs through history, but it is just possible that being hungry, dirty and ill produces the same effects.

Jeremy Bentham tried to establish a National Charity Company, a mixture of public and private poor relief, which offered food, clothing and employment but at the expense of being ruled and controlled. The poor would not be free under this control to act or make decisions for themselves. Many other reforms were suggested, tried, rejected – all indicated a growing problem which would not begin to be solved for another century.

Wandesford House, Bootham

5
York in the Seventeenth and Eighteenth Centuries

In the north, York was worst affected by the suppression of the religious institutions. Fewer than half of its medieval hospitals survived the sixteenth century. Those which had been closely tied to religious observance such as praying for benefactors or granting indulgences were condemned. The survivors were the almshouses and *maisons dieu* of the craft guilds, and those which had been handed over to the corporation, or had separate private benefactions.

There were eight new almshouse foundations in York in the seventeenth century, and one in the village of Heslington some two miles away. As in the medieval period, women were often responsible for generous benefactions, even if they were not publicly credited with them. Three of the new foundations were established by women: Anne Middleton's, Ann Wright's and Barstow's; and there were regular small benefactions given for the support of the poor living in them.

There was also civil war, and a siege which caused damage and destruction, but which ended less painfully than it might have, partly owing to the Fairfax family's relationship with the city. Sir Thomas prevented the sacking of York's churches and the Minster, and also saved its glass. There were others who kept York from destruction, including Edward Bowles who was chaplain to the forces under Fairfax, and was appointed as one of four ministers at York Minster. As a scholar he took a hand in saving both the Minster and its library from damage. He also (unsuccessfully) petitioned Parliament to establish a university in York; and he subsequently wrote a catechism, knowledge of which became a prerequisite of appointment to Lady Hewley's Hospital.

Civil War and York almshouses

In 1644, during the Civil War, the city was besieged by Parliamentarian forces. There were several batteries on high ground around the city, including one on Lamel Hill and another lower down in St Lawrence's churchyard, from which Walmgate Bar and Lawrence Street were bombarded, and the parish church formerly attached to St Nicholas's Hospital in Lawrence Street was destroyed. There was fighting and destruction in all the suburbs, where most of the houses were set on fire or pulled down to enable the guns to fire on the city walls. In Bootham Sir Arthur Ingram's Hospital, built only 12 years before, was wrecked. Gillygate and the Horsefair were similarly afflicted and what remained of the old hospitals was destroyed. In Monkgate, Agar's Hospital was burned down along with other houses in the street.

The area between Micklegate Bar and the fort on the Mount remained in the hands of the Royalists, and was the route taken by the wounded fleeing from Marston Moor. St Thomas's Hospital which stood immediately outside the Bar was perhaps well-enough protected but St Catherine's just below the Mount was in an isolated and rather vulnerable position. It was reportedly in a ruinous condition in 1648. No doubt during the siege it was deemed appropriate to move the occupants of the hospital to safety so it may, therefore, have decayed simply from being empty for several years. It may equally have suffered from the friendly depredations of Royalist soldiery – though at the end of the siege, the Parliamentarians took this route to enter the city through Micklegate Bar. The hospital's relative isolation can be seen in William Lodge's drawing of the city from the Mount showing the hospital in the middle ground to the left of the road. St Thomas's can also be seen to the right of the Bar.

***The Ancient and Loyall City of York,* 1678, by William Lodge**
showing St Catherine's Hospital from the Mount
(York Museums Trust (Art Gallery) YORAG:1988)

It seems not to have been recorded what happened to the occupants of any of these hospitals during the siege, though it is known that Ingram's residents continued to be paid and were able to return after it was rebuilt. There must have been some doubling up elsewhere, with other almshouses perhaps taking them in. An account of what happened to a group of Roman Catholic nuns living in Heworth Hall (the Old Manor House, now demolished) is likely to parallel what others did. Wenham (1970) includes the account, not only of their move into the safety of the city, but also of the mess they found when they returned. Mary Ward, foundress of the Institute of the Blessed Virgin Mary, is said to have died there shortly afterwards in 1645.

Eighteenth-century York almshouses

York's population grew from about 12,000 to about 16,000 during the eighteenth century, great numbers of whom were in poverty. A window-tax return of 1701 covering eighteen parishes, showed that the number of potential exemptions due to poverty, may have been as high as one third of the population in parts of the city. The Corporation provided some poor relief in kind as well as money. In the 1780s a fund was set up to provide coal and bread to the poor during a big freeze.

This was a period when wealthy women once again played a great role in charitable giving. Of the six hospital and almshouse foundations of the early eighteenth century, four were founded by women, and three of those still survive. Mary Wandesford's, founded in 1725, is still in its original house in Bootham. Dorothy Wilson established a hospital and school in her own house on Foss Bridge in 1717. Currently homeless, it plans to build a new almshouse. Lady Hewley's was established in Tanner Row in 1700 and still functions in St Saviourgate. Margaret Mason's was founded in 1732 in Colliergate but no longer survives as an almshouse. Thomas Colton established a Hospital at the bottom of Tanner Row in 1717 which survives in Shipton Street; and Robert Winterscales' Hospital was established in 1723, but no longer survives.

Several generous bequests by wealthy benefactors enabled these charities to continue to pay small stipends to their almspeople; other such bequests enabled hospitals for the sick and a dispensary in St Andrewgate to offer some medical services free to those unable to pay for them, thus continuing another Christian tradition.

There were also poorhouses including two in Acomb and Bishopthorpe, outside York. These were publicly-supported establishments, not almshouses as such, but are described in Part II as examples of the genre.

Lady Hewley's Hospital, Tanner Row
Drawing by George Nicholson, 1825 (*Evelyn Collection*)

York almshouses founded or rebuilt in the seventeenth and eighteenth centuries

The numbers refer to locations on maps 1-3 (p.69-71) and descriptions in Part II.

(16) Sir Thomas Hesketh's Hospital. 1608, rebuilt 1795. Heslington
(17) Watter's Hospital 1609. Nowtegail (became George Street, off Walmgate). Rebuilt in Chapel Row 1844.
(18) Agar's Hospital 1631 Monkgate.
(19) Ingram's Hospital 1630-32, rebuilt 1649. Bootham
(20) Anne Middleton's Hospital 1659. Rebuilt 1771 and 1828. Skeldergate.
(21) St Crux Almshouses 1663 probably in a court off Fossgate.
(22) Ann Wright's Hospital 1670. Bishophill. Location not known.
(23) Poor's Cottages 1691. Front Street, Acomb
(24) Sir Henry Thompson's Hospital 1692. Castlegate.
(25) Barstow's Hospital early eighteenth century. Blossom Street.
(26) Lady Hewley's Hospital 1700. Tanner Row.
(27) Dorothy Wilson's Hospital 1717, rebuilt 1765 and 1812. Foss Bridge.
(28) Colton's Hospital 1717. Tanner Row-Rougier Street junction.
(29) Winterscale's Hospital 1723. Walmgate.
(30) Wandesford's Hospital founded 1725. Built 1739. Bootham.
(31) Margaret Mason's Hospital 1732. No.21, Colliergate.

Medieval hospitals known to be still functioning in the seventeenth and eighteenth centuries

(6) St Catherine's Hospital 1333. The Mount. Rebuilt 1652.
(7) Holy Trinity Hospital c1371-3. Fossgate. Undercroft of the Merchant Adventurers' Hall.
(8) St Thomas's Hospital 1390. Next to Micklegate Bar.
(9) St Helen's fl. 1399. Fishergate. For lepers. Demolished 1622.
(10) Confraternity of St John the Baptist, ***Maison Dieu*** c1415. Aldwark. Rebuilt by the Merchant Taylors, 1730.
(12) Cordwainers' ***Maison dieu*** 1436. Near St Denys's church, off Walmgate.

6
Nineteenth-Century Charity: Housing the Poor in an Industrial Age

That the poor are with us always (Mark 14:7), is quoted often and perhaps rather complacently, as well as out of context, but the massive increases in poverty during the nineteenth century made it obvious that private charity and public poor laws could not meet the demand. The industrial age produced industrial-scale poverty, and levels of degradation almost beyond imagining. The problem of housing the poor was overwhelming, particularly as it affected public health. Mutual and Friendly societies, and groups of philanthropists working together began to address the situation, but it was only much later that radical state intervention began to make real inroads into solving the problem. In the meantime, the question of corruption among charities generally needed to be addressed.

Charity regulation

There had been several investigations into the workings of charities since the seventeenth century, but it was only in the nineteenth century that detailed surveys were carried out and abuses properly and successfully reported. The Brougham Commission investigated charitable trusts throughout England and Wales between 1818 and 1837 and recovered about £13 million in misused or under-used assets. Four hundred charities were referred to the Attorney General for prosecution, including the famous case of St Cross Hospital, Winchester which inspired Anthony Trollope's *The warden* (see Hampshire VCH). The results of the Commission's work eventually led to the Charitable Trusts Act of 1853 and the establishment of a permanent Charity Commission to oversee the accounts and activities of registered charities. To register as a charity, or to make a gift to charity in order to gain tax exemption, the purposes of doing so needed to come under one or more of four headings: the relief of poverty, the advancement of education, the advancement of religion, and anything beneficial to the community (an *et cetera* covering any other charitable activity that helped the poor, but might incidentally benefit the rich). This was in order to control abuses, and to direct charitable funds towards appropriate recipients.

Poor Law 1834

The Poor Law Amendment Act of 1834 passed by a Whig government replaced all previous poor law legislation. It was based on three principles: Malthus's, that population increased faster than resources (and therefore needed to be checked); Ricardo's 'iron law of wages' – that wages tend to fall to the minimum level necessary to sustain life, thus also keeping the population down; and Bentham's notion that people would do what was pleasant, and would therefore prefer to claim relief rather than work. They seemed unwilling to acknowledge that these principles were fundamentally questionable, or indeed that they were an implicit rejection of the principle of charity contained in St Paul's first letter to the Corinthians (in the King James version, of course). The chief flaw in the entire system was the rejection of the principles of kindness and compassion.

The Act established a Poor Law Commission to oversee the operation of the parish poor law unions, which built the workhouses. Outdoor relief continued to be permitted until 1840 when the only relief the poor could obtain was inside the workhouse. A rigorous means test also meant yielding up any property to cover the costs of their relief. The workhouses became virtual prisons in which couples were separated from each other and from their children. The 'less eligibility' principle made the workhouse a deterrent, and introduced a penal regime designed to be worse than the worst job outside. As the majority of inmates were old or infirm – unable to work – it is hard to see what the deterrent was.

Unemployed rural workers were encouraged to seek work in the towns, so there was a flow of such people into towns where work did not necessarily exist for them. They were then forced to seek relief in the workhouses, compounding what was very quickly becoming a scandalously mismanaged system. People often had to move away from their own parish to find work; but there was great reluctance on the part of the new parish to support them if the work dried up, or they became ill and unable to work – perhaps after many years living in the new parish. This intolerable conflict lay at the heart of the system of parish-relief.

In the 1830s, some of the able-bodied poor were put to work repairing roads to earn their keep – it was said that the roads had never been in such good condition – but it must have had an effect on the wages of genuine road-menders.

Pauper children, often dirty, foul-mouthed, diseased or criminal, were an unwelcome group anywhere, including schools They needed support; they needed to be taught how to support themselves so as not to become a burden in adulthood; but not many people were prepared to suffer these particular children, little or otherwise. Those who were prepared to do so included the Quakers, who started schools for them which were working well by the end of the century. Elizabeth Fry, the Quaker reformer of prisons, even set up schools for children who lived in prison with their parents. Others established Ragged Schools. But there were also the immigration programmes to Australia and Canada, which sent thousands of children to what was supposed to be a better life, separated from friends and from family (not all were orphans, even if they were told they were). In fact it was often into hardship, if not slavery. It was a well-meaning, but essentially evil practice that lasted until 1967 and resulted in much misery.

Hatred of the workhouse and outdoor relief, and all the other trappings of what a caring society can impose, was the dominant feeling among the recipients. Often the feeling was mutual: Assistant Poor Law Commissioner Edward Gulson in Nottinghamshire saw no reason to allow any gifts of tea or tobacco to inmates of the workhouse, preferring to administer severe discipline undiluted with kindly benevolence. It is not surprising that paupers were often prepared to lead a hand-to-mouth existence, and some preferred to starve, in the hope of finding support before being forced to take up an "offer of the house".

The punitive character of the workhouse did not necessarily translate to almshouses because almspeople were chosen as worthy beneficiaries. Bill Sessions' notes on York almshouses a century later, however, suggest that the almspeople sometimes felt that their poverty did not endear them to the administrators, even though this may partly reflect the fact that recipients do not always perceive well-meaning acts in the way they were intended.

Public awareness

It is clear from legislation in what low esteem the poor were held by the state, but this was not necessarily the case with private individuals. There was considerable public opposition to the Poor Law Amendment Act, not only on financial grounds but also on moral grounds – Malthusian ideas of limiting childbirth were regarded as unnatural and obnoxious, splitting up families was intolerable, and locking up free Englishmen without due process was outrageous and deeply resented. There was strong local feeling against the new Poor Law in York and Yorkshire. The York Herald of 20 May 1837 reported a meeting with a Poor Law Commissioner held in the Merchants' Hall, which had to be moved to the Guildhall. The Commissioner's responses to questions were specious, and begged more questions than were answered. His audience seemed to think so too. A tea-dealer, Mr Jackson of New Market, asked a question about the separation of husbands and wives in the workhouses and stated, to applause, "I don't think it a crime to be poor, or that we have a right to grind them down," (a direct reference to Isaiah 3:15). He was not alone, as a report on the next page showed. An enormous meeting in the West Riding was attended by (apparently) 100,000 people to oppose the "obnoxious clauses" in the Amendment Act, particularly the creation of prison workhouses for free Englishmen, but also the separation of families. Banners were waved which read, for example, "No Bastiles [sic] for Englishmen", "Remember Heaven has an avenging rod. To smite the poor is Treason against God". The old biblical injunctions to aid the poor (and not grind their faces) were still strongly felt.

Our own knowledge of the human cost of extreme poverty comes in part from writers like John Clare who wrote tellingly from personal experience in his poem *The parish: a satire*. Charles Dickens' descriptions of raw poverty were, as he said himself, necessarily a good deal milder than the truth but his observations on destitution and his descriptions of London slums in *Bleak House, Oliver Twist*, and particularly *Sketches by Boz*, remain shocking because he was shocked by them. He also wrote in *Household words* in May 1850 "… we have come to this absurd, this dangerous, this monstrous pass, that the dishonest felon is, in respect of cleanliness, order, diet, and accommodation, better provided for, and taken care of, than the honest pauper…" which may not have been entirely true, but made its point.

Dickens' contemporary, Henry Mayhew, also described what he saw in London and revealed some of the horrors of living in poverty in a big city. He looked at the poor in a fairly traditional way, in terms of "those that will work, those that cannot work, those that will not work" and described their lives accordingly. He also looked at the city of dreadful night, London's underworld, the remedy for which seemed to be beyond any ordinary charitable endeavour, but which posed a serious health risk and needed some kind of solution. Another contemporary, Charles Kingsley, questioned society's treatment of the poor, and particularly of children, in *The water babies*. At the end of the century there was also Thomas Hardy whose description, in *Far from the madding crowd*, of a woman in labour trying to get to the workhouse is intentionally distressing reading. These authors deliberately evoked shocking images to alert their peers to a social evil which required remedy, not unthinking condemnation of helpless individuals caught in circumstances beyond their control.

More pertinently, Charles Booth established the ground rules by which judgements about the nature of poverty could be made. He established a notional idea of a poverty line, and by use of statistical information and his mapping of poverty in London in the 1890s, proved (to his own surprise) that what had seemed exaggerated figures for poverty levels were considerable underestimates. The number of people below the poverty line was not a quarter of the population of the city, but more than a third. Fearful of socialist revolution he argued for the introduction of old age pensions.

Among philanthropic responses there were, as ever, the Quakers whose concept of charity included compassion and kindness and who developed practical responses to social deprivation, not only schools but training of adults to enable them to help themselves out of their condition. They started night-time shelters for people sleeping on the streets in winter and after Elizabeth Fry's death in 1845 even established refuges in her memory. This was akin to reinventing the wheel as it was how almshouses and hospitals started in the first place. Later in the century the Salvation Army, with similar principles, started to offer practical support to the poor and destitute, and they too were repeating the kind of work first recorded in England more than nine centuries before.

In London, the condition of the poor led one wealthy benefactor, George Peabody, an American, to give the staggering sum of £500,000 in 1862 to set up the Peabody Trust to build tenement blocks with a good water supply for the deserving poor. But even he could not help the utterly destitute because they had no money to pay the small rent. Others set up voluntary housing organisations and model-dwelling companies to create affordable housing in areas where land values were high. The buildings were often high-rise, high-density tenements, intended to combine good standards of building and, crucially, good sanitation. These companies raised share capital on the promise of a guaranteed five-percent return. Charitable trusts invested the endowments of rich philanthropists who were prepared to accept a lower rate of return. Even among relatively high-minded investors the temptation was to create high-density housing, for maximum profitability, often at the expense of the health and well-being of the poor who had to live there. The link between high investment returns and inadequate provision has always been strong. But London with its enormous population was a special case and little of even this level of philanthropy occurred in the provinces. Titus Salt's efforts to improve the conditions of his own workers in Bradford were a beacon of light amid the horrors of the time.

Tablet from the original foundation of Lady Hewley's, including her arms, reset into the wall facing the street:
"This Hospital was founded and endowed by Dame Sarah the relict of Sir John Hewley of the City of York Knight, Anno Domini 1700. Thou O God hast prepared of thy goodness for the poor. Ps 68.10"

7
Nineteenth-Century York

Conditions in the slums

Their size gave London and the Coketowns of the north profound problems with poverty, but York's slums were bad enough. There were pockets of utter degradation, and mortality rates, particularly of infants, were higher than in most other towns. Accounts of the 1832 cholera outbreak in York list the streets where people had fallen sick – nearly all were slums, and include Aldwark, Bedern, Friargate, Mucky Peg Lane (now Finkle Street), North Street, Paver Lane, Skeldergate, Shambles, the Water Lanes, Walmgate, and many more. Thousands of people lived in these slums, which were crowded, infested and crime-ridden. It may have been possible to ignore the existence of these areas, but disease and crime have a tendency to spread beyond borders and demand attention. Sufferers from epidemic disease passed it on to their families in other areas, and indeed other towns, via their clothing and bedding.

Poor-quality, damp, crowded housing meant that the diseases of poverty: tuberculosis, pneumonia, typhus, typhoid, as well as cholera and the many childhood diseases, took hold very readily. Timber-framed houses, picturesque now, once harboured a variety of less picturesque vermin. When Irish refugees fleeing the famine of 1845 came to York, they brought typhus with them which, carried by lice, spread quickly among the tight-packed slum-dwellers. Its other name is gaol fever. The coming of the railways brought the navvies, who came also mainly from Ireland, building the track and following the progress of the railways round the country. In York they congregated in Bedern and Walmgate and the narrow lanes and courtyards of other cheap slum areas of the city where there was serious deprivation (as well as unruliness and violence) and not much sign of alleviation. In Back Bedern, for example, five privies were shared by three hundred people. There were many areas whose occupants relied solely on a midden. There was little infrastructure of drains and sewers; instead there were open drains, uncollected refuse, human and animal waste, and in the Shambles there were also slaughterhouses and piggeries. It must have been far worse than "the abominable smell abounding in the city" complained of by Edward III in 1332. Money set aside for drainage and sewerage tended to be spent in the wealthier and more influential areas of the city rather than in the slums where it was most needed.

In 1844 Dr Thomas Laycock was Physician to the York Dispensary and also taught at the former York Medical School in the County Hospital. He later became a professor in Edinburgh, specialising in neurophysiology. He had, however, studied statistics in Paris, and was able to provide a grim report to the Commissioners for Inquiring into the State of Large Towns and Populous Districts. He found that there was a link between proximity to the river (and the water-table) and early death. It was proved by comparing mortality statistics of different parts of the city, as well as of villages nearby. People tended to live longer in Acomb than in Rufforth, for example. The low-lying areas of York included many slums like those in Walmgate, Hungate and North Street, and mortality statistics reflect the fact. For the labouring classes living in the parish of All Saints North Street the mean age at death was less than twenty years. This was the result of living at the foot of a hill near the river where, particularly in wet weather, the overflow from every cesspit and midden-privy above came through the houses. Higher up the slope of Micklegate, in the parish of Holy Trinity, the mean age at death was forty-three. It is interesting in this context to note that, during the 1854 cholera outbreak in London, the doctor responsible for finding a statistical link to the Broad Street pump was one John Snow (1813-1858), who was actually born in North Street, York. The commemorative plaque to him on the Park Inn hotel is well hidden. But even Laycock's appalling statistics for York represent better conditions than those prevailing in the large industrial towns of the north, like Bradford, where only thirty percent of textile-workers' children achieved the age of fifteen, and those that did might not live beyond eighteen.

After 1850 the York Board of Health began to attempt some sort of improvement of the environment. Despite all efforts, they and the City Commissioners failed to keep the problem of human and animal waste under control. In 1886 there were still eight thousand midden-privies, and no systematic disposal of the contents. In addition, while local authorities were empowered to demolish slums, they had neither the money nor the power to replace the lost housing stock. Whole streets were demolished in slum clearances, giving timber-framed houses in the Water Lanes, for example, no chance to exhibit their picturesque qualities in better times; the inhabitants had nowhere to go but to double up with someone else. Whole families might occupy one room, and the problem of overcrowded tenements emerged again elsewhere. Laycock recorded many cases of large families occupying one room and, in at least one case, lacking bed or bedclothes they slept on a pile of shavings in one corner. In the former Bedern College buildings sixty-seven families had each only one room. Charles Dickens reported as many as three families occupying a single room in some of London's slums.

There was charitable distribution to the poor of food, fuel and money, but looming over them were the workhouses which both stigmatised and demoralised the people they purported to help – as grudged charity often does. The Marygate Workhouse, established in 1767 and housing about ninety paupers, was by the nineteenth century in an indescribably squalid state. That children there were suffering from venereal disease gives some inkling of the conditions. A new workhouse was built in Huntington Road in 1848-9 to take more than six hundred paupers, who received some rough clothing and a bed. (It has recently been refurbished and turned into desirable flats.)

Charities in York: the nineteenth century

By the nineteenth century York was extremely well-supplied with charitable endowments. The Brougham Commission examined and reported on around two hundred York charities between 1820 and 1828, whose endowments produced an annual income worth more than £5,000. Forty percent of this, some £2,000, was devoted to almshouses and associated pensions. A slightly smaller amount, about £1,500, was disbursed for the poor in other ways; a further £1,130 supported children's education and apprenticeships; and the rest was used for the support of ministers and churches. At the end of the century there were nearly three hundred endowments whose gross income was more than £8,700. A smaller proportion of that sum was devoted to almshouses – about a quarter, but more was appropriated for education and apprenticeships, a little under forty percent. The poor and the churches received about half of the remainder each (VCH 1961).

The charities were controlled by different groups. There were many independent charities, which included some of the larger almshouses and the York Dispensary Sick Poor Fund, and these controlled the bulk of charity money (by 1939 this amounted to around sixty-five percent). About a quarter of charity income was controlled by parochial charities, some of which were very poor, but some of which were quite wealthy from their ownership of land and property. The rest – some nine percent – was under the direction of the Mayor and Commonalty, the body that had taken over almshouses and *maisons dieu* in past centuries. Legislation in 1837 removed the charities from their direct control and put them instead into the hands of thirteen York Charity Trustees. These trustees ran all the municipal almshouses, listed in Table 1 (p.127), whose individual fates were decided in the twentieth century.

York almshouses founded in the nineteenth century

Several almshouses were founded or rebuilt in York in the nineteenth century making a total of about twenty-six by the end of the century. The older houses tried to continue to support the deserving poor, but in conditions which were deteriorating. Some drifted into decay and dereliction for lack of funds and possibly lack of interest on the part of trustees (and therefore potential benefactors).

Two medieval foundations were demolished and rebuilt elsewhere in this period: St Catherine's was moved from the Mount to Holgate Road in 1835; and St Thomas's Hospital was demolished in 1863 for the widening of Nunnery Lane and moved from outside Micklegate Bar to a position next to Victoria Bar in Nunnery Lane.

Most almshouse provision was for poor elderly widows, or spinsters as at Mary Wandesford's. Some benefactors were more generous than others and concerned for the well-being of their almspeople. This was certainly true for people living in Lady Hewley's and Dr Colton's hospitals, where maintenance was carried out regularly, and in the case of the former, piped water was supplied to the original almshouse in Tanner Row to make life easier.

Very few offered space to men or couples. Sir Henry Thompson's was for men only; the Merchant Taylors' and Sir Thomas Hesketh's were mainly for men. Lady Hewley's later almshouse in St Saviourgate had space for couples.

The nineteenth-century foundations were mostly quite small, and some lasted a relatively short time compared with earlier foundations. Briggs' Charity lasted about a century, moving several times in the area of King Street and Low Ousegate. Barstow's Hospital in Blossom Street was demolished in about 1859, and the inmates moved to new accommodation at the Clementhorpe end of Nunnery Lane. The cottages were still occupied when Elizabeth Brunskill described them in 1960, but were demolished shortly afterwards. Barstow's did survive as an almshouse by merging with the twentieth-century Cameron Walker Homes in Bishopthorpe Road. Sister Wilson's lasted just over seventy years in East Parade, Heworth, before being demolished. Sir Joseph Terry Memorial Homes, built in front of Anne Middleton's were founded in 1899, and survived as an almshouse until the 1970s when it became part of Lady Anne Middleton's Hotel. Surviving establishments of this period are Lady Hewley's, Ann Harrison's, and Ellen Wilson's.

Blossom Street, looking towards Micklegate Bar
Coloured lithograph by William Monkhouse, c1845, showing St Thomas's Hospital to the right of Micklegate Bar, and Barstow's Hospital, the timber-framed building on the extreme left
(Evelyn Collection; York Museums Trust (Art Gallery) YORAG:R1776)

York almshouses founded or rebuilt in the nineteenth century

The numbers refer to locations on maps 1-3 (p.69-71) and descriptions in Part II.

(12) Cordwainers' *Maison dieu.* Originally built 1436. Rebuilt 1811. Church Lane, off Walmgate.
(17) Watter's 1609. Newtegail Lane, (later George Street, off Walmgate). Rebuilt in 1844 in Chapel Row
(20) Anne Middleton's 1659. Rebuilt 1771 and 1828. Skeldergate.
(27) Dorothy Wilson's 1717, rebuilt 1765 and 1812. Foss Bridge.
(32) Briggs' Charity 1802. King Street; moved to Low Ousegate 1860s; defunct before 1907.
(33) St Martin's Cottages pre-1824. St Martin's churchyard, Coney Street.
(34) St Catherine's Originally 1333 The Mount **(6)**. Rebuilt 1835 in Holgate Road.
(35) Lady Hewley's Originally 1700 Tanner Row **(26)**. Rebuilt 1840, St Saviourgate.
(36) Ann Harrison's 1845. Penley's Grove Street/St John Street.
(37) Bishopthorpe Poorhouse 1846. Main Street, Bishopthorpe.
(38) Barstow's Originally Blossom Street **(25)**. Rebuilt 1859, on the corner of Caroline Street and Clement Street at the Clementhorpe end of Nunnery Lane.
(39) St Thomas's Originally outside Micklegate Bar **(8)**.
Rebuilt 1863 next to Victoria Bar, Nunnery Lane.
(40) Sister Wilson's 1885. East Parade, Heworth.
(41) Ellen Wilson's 1894. Lawrence Street, next to St Lawrence's church.
(42) Terry Memorial Homes 1899. Skeldergate.

Earlier hospitals still functioning in the nineteenth century

(7) Holy Trinity/ Merchant Adventurers Hospital c1371-3 Fossgate. Undercroft of the Hall.
(10) Merchant Taylors' *Maison dieu* rebuilt 1730. Aldwark, formerly the Confraternity of St John the Baptist.
(16) Sir Thomas Hesketh's Hospital. 1608, rebuilt 1795. Heslington.
(18) Agar's Hospital 1631 Monkgate.
(19) Ingram's Hospital 1630-32, rebuilt 1649. Bootham.
(21) St Crux Almshouses 1663 probably in a court off Fossgate. Transferred 1895 to Piccadilly.
(22) Ann Wright's Hospital 1670. Bishophill. Location unknown. Charity consolidated with others 1907.

(23) Poor's Cottages 1691. Front Street, Acomb.
(24) Sir Henry Thompson's Hospital 1692. Castlegate.
(27) Dorothy Wilson's Hospital 1717, 1765, 1812. Foss Bridge.
(28) Colton's Hospital 1717. Tanner Row.
(29) Winterscale's Hospital 1723. Walmgate.
(30) Wandesford's Hospital 1725/1739. Bootham.
(31) Margaret Mason's Hospital 1732. Colliergate.

St Thomas's 1862 hospital at Victoria Bar

8
Twentieth-Century Responses to Poverty

Benjamin Seebohm Rowntree, a Quaker, carried out three surveys of poverty in York between 1899 and 1951, during which time conditions did begin to improve. He defined the poverty line as a minimum weekly sum "necessary to enable families... to secure the necessaries of a healthy life". He also looked, more significantly, at nutritional needs and the cost of achieving a healthy diet, and determined that nearly twenty-eight percent of York's population lived below the poverty line – a similar figure to Charles Booth's figures for London. Those in old age or extreme youth were most likely to be in abject poverty, through no fault of their own. It is relevant to note here that a feature of York's population in the 1970s, obvious and rather shocking to this newcomer who had never seen it before, was the number of elderly people, women in particular, of very short stature who had bandy legs. These were clear indications of the rickets and malnutrition that these people had suffered as poor children at the turn of the twentieth century and in the Depression years.

Rowntree demonstrated that poverty was the result of low pay, and that the poor were not necessarily responsible for their own situation, even though some were spending their otherwise-sufficient income unwisely. The effect of these and other surveys led to state intervention in the task of alleviating poverty at the turn of the twentieth century, including the introduction of the old age pension. However, a follow-up study of the families in the Rowntree reports (Atkinson 1983) carried out in the early 1980s, suggested that, in sociology-speak, intergenerational income mobility had not really changed very much. In other words, families who started poor tended to remain poor. This was further confirmed by a survey carried out to mark the centenary of the first Rowntree report on York (Huby, 1998) a study which details the problems still encountered by the poor in York.

Old age pensions

The Old Age Pensions Act 1908 introduced the first state pension. It was a non-contributory pension designed to give a minimum standard of living to the 'aged deserving poor' who would otherwise be destitute. At that time, there were approximately two million people over sixty-five in Britain, of whom two thirds were estimated to be in want, though not necessarily eligible for poor law relief, and who were 'for the most part simply suffering from the sin of living on' (Salter 2009). This pension was given only to people of good character, in need and over seventy (similar criteria to those for charity provision over the centuries). It did not provide enough to live on, merely a safety net. A parallel means-tested income ran alongside for the very poor – which incidentally further stigmatised them. Richard Titmuss, a pioneer in social policy, remarked that there existed "two nations in old age" – those with comfortable provision in old age, those without. The poor still had to work well into old age if they were to keep fed, clothed and housed.

The first contributory state pension scheme was introduced in 1925, providing an entitlement from age sixty-five. The retirement age for women was lowered in 1940 to sixty; but women workers, being largely in low paid, part-time or temporary work, lost out on significant pension benefits. It is still the case that women remain most likely to be poor and in need of support. In 1948, a pay-as-you-go state pension scheme was introduced, with current benefits paid out of current contributions, which was intended to provide an income just sufficient to live on.

Social change after 1947

Social and economic change since 1947 has altered the nature of poverty in British society. The old poor laws were abolished by the Attlee government's National Assistance Act 1948. The state pension, and associated benefits for the poor, have (it seems) caused the almost unimaginable poverty that existed before the twentieth century practically to disappear. However, twenty years after the welfare state came into existence, Jeremy Sandford wrote his television play *Cathy come home*. Watched by twelve million people, the play was about the sudden unemployment of a father, caused by an injury, and his subsequent destitution. Homeless, his family is split up and the children taken away from their mother. The play was a peculiarly shocking revelation of the realities of the housing crisis and how poverty can descend without warning on a blameless victim. It had an enormous effect on public opinion, suddenly provided with the means to make the imaginative leap into the lives of the poor. The charity Shelter gained prominence as a result of this one play, and moved from being a small local organisation to working on a national basis. It had been founded in the same year, 1966, by the Rev. Bruce Kenrick who had earlier established the Notting Hill Housing Trust to provide decent affordable houses for rent. But it took another eleven years before local authorities became obliged to house the homeless.

However, academic and government surveys published at intervals since then indicate that poverty, however it is defined, is still more common in old age. Alan Walker reported, in *Ageing and society* in 1981, that one in four of the elderly had incomes equal to or below the poverty line, compared to one in twenty of non-elderly people. This was a moderate improvement on the two-thirds of the elderly poor living in or on the margins of poverty in 1908. In 2000 a House of Commons Select Committee Survey on Social Security, called *Pensioner poverty* (Seventh report), suggested that while pensioner incomes rose by seventy percent between 1979 and 1997, these figures concealed "serious and growing income inequalities among pensioner households".

In 2009 the Work and Pensions Committee reported that even after a one-third drop in the number of pensioners living in poverty, nevertheless there were two million still in want. One in five pensioners was classed as living in relative poverty. National Office for Statistics population figures for 2008 put the number of people over sixty-five at 11.8 million - between five and six times what it was in 1908, and higher than the number of under-sixteens (11.5 million). If twenty percent are in relative poverty, this actually represents some 2.36 million people.

The baby boom generation

The Nuffield survey of poverty and the care of the elderly, *Old people*, was published in 1947, when provision of support for the poor and the elderly was about to change utterly. The welfare state, created after World War II, gave state pensions to all the old, paid for out of the contributions of the working population. Since then other forms of social and financial support have been provided as part of cradle-to-grave welfare. The potential for this to become a burden was raised in *Old people*, where it was pointed out that the population was aging and there might not be enough people of working age to support them after 1989. Its population projection was for three million fewer working people in 1989 than there were in 1946, a projection which was made redundant with the homecoming of demobbed servicemen and women.

The unforeseen postwar baby boom began in 1947, a year with the biggest birth count on record, but we have come full circle. In 2012, 800,000 people will turn sixty-five (Hargreaves, 2010). It remains to be seen whether state pensions and other benefits, paid out of current contributions (in effect a legalised 'Ponzi' scheme) will be able to support the aging baby-boom generation because pensioners may soon outnumber the working population, as they already outnumber the under-sixteens.

According to Tony Salter's *100 years of state pension*, published in 2009, the proportion of people over sixty-five represents a thirty percent dependency ratio of pensioners to the working population which, with increasing life expectancy, is why the retirement age is to rise. Others suggest the ratio of workers to retired is 4:1. Forecasts for the next forty years to 2051 predict an increase in the dependency ratio to 2:1, which would place a crushing burden on the younger population. The cost of social care of the elderly is expected to rise to £24 billion by 2026, according to the volunteering charity Community Service Volunteers (Third Sector online journal, 7 July 2009). But, as in 1947, such forecasts are vulnerable to real events and, as such, might change.

Public and private charity in the future

The future is uncertain. There is enormous change in the economic as well as the global climate. There has been a political change in the UK. All of these will affect the poor and the old, because they always do. There will be measures to cut public spending, a period of austerity, and the vulnerable may find they have to justify their need of support. In the sixteenth century, the collapse of monastic provision of services to the poor and old stimulated private benefactors to give enormous sums and endowments to replace that provision. Some of that money came from not necessarily honourable sources – in York, for instance, Sir Arthur Ingram was an unscrupulous financier and land speculator in addition to being a great benefactor. Others derived their money from investment and industry: some of Mary Wandesford's money came from investments in South Sea stock; the Hewley wealth derived from ironstone mines; much wealth depended on the value of land.

Twenty-first-century wealth is similarly derived, but the balance lies with money made from money, and less from industry or land. In 2010 the Sunday Times rich list showed that the thousand richest people increased their wealth by £77 billion to a total of £335.5 billion, or one third of the national debt. There are fifty-three billionaires in Britain. Some of their wealth comes from industry and property, as well as inheritance, but a great many of these billionaires are City bankers, financiers, and hedge fund managers, who keep a higher proportion of their already huge income than the poorest fifth of the population keep of their tiny income. No doubt many contribute to efforts to improve such global problems as starvation, disease, and climate change. Will it be considered it a moral duty to help the poor at home in a new age of austerity?

Individual charities' claims to offer a public benefit are being more closely examined by the Charity Commission. As in the past, some charitable organisations become less relevant as a result of social change, so this is a healthy procedure, and not necessarily a sign of corruption. Charities are exempt from some (a few) kinds of tax. For a fair offset they need to remain free enough of taxation to ensure their money is spent on areas that the state cannot reach.

Burrill Homes, Water End, Clifton
Decorative Carving

9
York Almshouses in the Twentieth Century

Almshouses 1900-1946

In 1909 there were 143 almspeople in York, and a Royal Commission report on the Poor Laws and Relief of Distress (Kay, 1909) described the kind of circumstances these individuals had found themselves in before entering an almshouse. The Commission also found that, though the municipal charities were behaving in a properly transparent fashion over the selection of beneficiaries, the privately endowed ones were not. There was a danger of their trustees appointing only protégés. The Commission also found that endowments were increasingly unable to keep up the stipends to almspeople, who needed consequently to be helped by the Guardians of the Poor. The municipal charities consisted of St Catherine's, St Thomas's, Middleton's, Thompson's, the Cordwainers', Sister Wilson's and Terry Memorial Homes. Exemplary in their selection policies they may have been, but none of these has survived as an independent establishment.

In the 1930s two new almshouses were established in Clifton, John Burrill Homes on Water End, and Fothergil Homes in Avenue Road, both designed by architects Ward and Leckenby, and both, for the first time, offering internal bathrooms with hot and cold running water, a hitherto unheard-of luxury in almshouse accommodation.

Conditions

That the conditions of almshouses had begun to deteriorate had been apparent for some considerable time, in some cases since the nineteenth century. William Hargrove's *History* of 1818, and Baines' *Gazetteer* of 1823 both describe conditions which were poor even by the standards of their day. Some had to wait until the twentieth century for improvements to be made, or for demolition if they could not. Those almshouses lucky enough to be moved in the nineteenth century, such as Lady Hewley's, provided and were able to maintain better conditions.

In the period after World War I, and even more after World War II, when the country was bankrupt and there were people needing urgently to be housed, there was no money for sensitive, historically-in-keeping or, sometimes, even necessary repairs and maintenance. York was lucky in having relatively little war damage, but even so a great many buildings were destroyed and people killed. Bill Sessions remembers coming home from working in blitzed London to find central York looking neglected, even blighted. He was sad to see that even the Shambles seemed beyond repair, while the roof of the Merchant Taylors' Hall had gaping holes and serious rain damage. In the economic conditions, the trustees and clerks of almshouses were very often not able to afford to put things right.

Many continued to pay their sometimes very poor inmates a stipend, but the Depression and the two world wars undoubtedly reduced the income available for financial support of inmates and for maintenance of the buildings. As a result there was decline, and some fell into the dereliction reported by Bill Sessions in Appendix 13 of *Old people*. It is clear from his notes that some of the almshouses surviving to 1946 had become inconvenient, insanitary, dirty, damp, and very unhealthy. Almspeople also shared sinks, cold taps and outside WCs, and usually lacked a bath, shared or otherwise. Slopping out must have been the norm, and often buckets of slops, water, or coal, had to be carried up and down stairs by the elderly. Of the approximately 174 almspeople, only seventeen had an individual bath; 135 had no bath, no washbasin and no access to hot running water. The houses were often in poor repair and had poor decoration: residents were mainly expected to pay for redecoration themselves. Half of the almshouses were lit by gas – which many preferred to the electric alternative – but there was often poor lighting on stairs. War damage had affected several of the newer almshouses which had happened to be in the way of the bombing, and while the damage was not as severe as what happened to Agar's and Ingram's Hospitals in the Civil War siege of 1644, was quite bad enough. However, the basic conditions described were not unique to charity property: many people lived in similar conditions well into the 1970s.

Almshouses and York Charities

In the early 1900s there were forty-five endowments in the hands of municipal trustees. Charity funds, including those previously intended for apprenticeships or the provision of loans, were becoming chiefly devoted (for lack of applicants) to the support of aging women, particularly widows. Occasional money doles were converted into regular pensions or gifts in kind. Other charity relief in kind was converted into provision for nursing and convalescent homes and hospitals, to complement welfare services provided by the state. In 1910 a Charity Commission bill to consolidate appropriate endowments and thereby direct charitable activities in some purposeful way was notoriously rejected by York's many charity trustees, which caused the measure to be dropped. These included private and parochial charities which all seem to have been jealous of their small empires. A more successful bill was passed in 1956 amalgamating the municipal charities only.

Nine of York's twenty almshouses were directed by York Municipal Charities (now York City Charities) and administered by solicitors Crombie Wilkinson. In the late 1940s Lewis Waddilove, an energetic secretary of the Joseph Rowntree Village Trust, became a trustee. He and Charles Robinson, a partner with Crombie Wilkinson, surveyed their almshouses and, given the state of some of them, drafted a proposal to sell the buildings and merge the charities.

The plan was agreed with the Charity Commission and led to a private Act of Parliament (4&5 Eliz.2 c.53) unhelpfully entitled "Frances Barker and certain other Charities (City of York) Scheme Confirmation Act, 1956", which merged a total of forty-six charities. None of the almshouses listed under this Act now survives under its own name, though several of the original buildings survive. All have been amalgamated as York City Charities and are administered from Thomas Fothergil Homes. Listed in the Act were the municipal charities mentioned above: St Catherine's Hospital; Thomas Fothergill [sic] Homes; Sir Arthur Ingram's Hospital; Mrs Ann [sic] Middleton's Hospital; Terry Memorial Homes; St Thomas's Hospital; Sister Wilson's Home; and Sir Robert Watter's Hospital. It also included the Cordwainers' Pension Charity, the Charity of Lady Hewley (which was separate from the almshouse charity), Sir Henry Thompson's Pension Charity, and the Charity known as the Cremitt Money (an Elizabethan charity whose origins date from the dissolution of St Leonard's Hospital in 1540).

The Act specifically permitted two almshouses on the list to be sold: Sir Robert Watter's and Sister Wilson's. Of the other almshouses whose buildings have gone since 1947, two were already listed for demolition: Sir Henry Thompson's in Castlegate and Robert Winterscale's in Walmgate. Sir Henry Thompson's had already been sold for site value and demolished in 1949. Sir Robert Watter's in Chapel Row was pulled down in the late 1950s. St Catherine's in Holgate Road survived as an almshouse until about 1962, when it was demolished; Sister Wilson's, founded in 1885 at 17-19 East Parade, Heworth, was neglected in the twentieth century and demolished in about 1957.

Of those whose buildings survive, several became hotels, reverting thereby to an original meaning of the word hospital. St Thomas's nineteenth-century establishment survived in Nunnery Lane until 1972, when it was sold and became the Moat Hotel. Anne Middleton's and the Joseph Terry Memorial Homes closed in the 1970s and both now form Lady Anne Middleton's Hotel.

Ingram's Hospital was bought by J.B. and Cuthbert Morrell in 1957, and was turned into five flats for private rent. There was a transfer of a piece of land between the back of Burton Croft (John Bowes Morrell's home) and the Fothergil site, which became part of Fothergil Homes to house the former Ingram's Hospital residents.

Crombie Wilkinson are still responsible for running Fothergil Homes and York City Charities; its clerks have included Roger Wilkinson, Anthony Royle, and Charles Robinson. The present clerk, Richard Watson, is currently Sheriff of York (2010).

Some independent almshouses have also gone. Margaret Mason's at 21 Colliergate closed in 1958 and is now a shop; the Merchant Taylors' *maison dieu,* rebuilt in 1730, survived until the 1940s after which it was converted into a single cottage for the caretaker of the Taylors' Hall. As noted above, Barstow's off Nunnery Lane was pulled down in the 1960s, though the charity merged with Cameron Walker Homes and still survives.

With the closure of so many, it may have seemed that almshouses were an anachronism too far. However, twelve almshouses in total remain (there is also Dorothy Wilson's, currently homeless, but intending to build a new almshouse). Together with Fothergil Homes, these are: Burrill Homes on Water End, Cameron Walker Homes (now expanded to become Walker Barstow) in Bishopthorpe Road, Colton's Hospital in Shipton street, Ann Harrison's Almshouses, rebuilt in 1964 on the corner of Penley's Grove Street and Abbot Street, Lady Hewley's Cottages in St Saviourgate, Wandesford House in Bootham, and Ellen Wilson's Hospital in Lawrence Street. In addition, Heslington Almshouses Charity manages two in Heslington: Sir Thomas Hesketh's Hospital and Lady Deramore Memorial Almshouses.

The list also includes the two new almshouses that have been established since 1947. One is in Fulford, Sir John Hunt Homes, built in the early 1950s on a large open site on the corner of Fulford Road and St Oswald's Road. The other is an almshouse established by Jane Wright's Charity in Ogleforth and Saville Court in 1983, which has become one of York's minor and unexpected pleasures – a small enclosed area of quiet inside the walled city.

Five of the almshouses are open only to women: Ann Harrison's, Colton's Hospital, Ellen Wilson's, Lady Deramore's, and Wandesford House.

Conditions after 1960

It has been for many years unthinkable that anyone should have to slop out or rely on shared outside privies and cold taps, a tin bath, or to be without hot running water – conditions obtaining in private homes as well as in many York almshouses in 1947 and for many years afterwards. In the last thirty or forty years there has been growing expectation of more comfort – hot water, central heating, a decent kitchen and bathroom, and timely repairs. The almshouses in York are now well able to compete with other types of accommodation for the vulnerable. They have to if they are to attract residents and keep them.

From the 1960s onwards, most almshouses in York had to start charging a maintenance contribution. Even those with endowments or investment income needed to do so. The minutes of Wandesford House, for instance, reveal a need in 1966 for help to pay for the cost of heating and hot water, but also to build up a reserve to pay for repairs. Indeed, when major alterations were to be carried out two years later, the advice of the National Almshouses Association was that before the Trustees could ask for grants or loans the residents should be making some sort of contribution. At that point contributions began to be levied. Other almshouses followed suit as circumstances dictated – some, like the Burrill Homes, only recently.

Government measures to provide income support, such as Housing Benefit and Council Tax Benefit mean that almshouses can charge their residents a realistic amount to live in them, which in turn pays for improvements and maintenance of the buildings.

Almshouses continue to be founded occasionally, but they are not the same institutions that they were two, three and four hundred years ago. They are no longer a major player in the alleviation of poverty. Their traditional charitable role remains that of enabling elderly people on reduced incomes, who might otherwise live in hardship, loneliness and distress, to live independently within a supportive community. What remains important is the independence of the provision of such help. Charities can work in cooperation with the state and at the same time promote a continuing awareness of social need.

Almshouses are an often-picturesque element in the streets not only of York, but also of many small towns and villages. Financial sustainability enables trustees to maintain them so that they can remain private spaces for the elderly, rather than historical fossils, while still providing an attractive architectural amenity.

There have been small but significant social changes within the almshouses too: one is the way in which residents are described and which reflects attitudes towards them on the part of the trustees. They were, for example, frequently referred to as “inmates” or, more genteelly, as “lady inmates” in the minutes of trustees' meetings at Wandesford House until the 1960s. This began to change to “the ladies”, and more recently has become “the residents”. This is the language of equality.

There is a certain amount of competition to find residents for different types of old-age homes, and almshouses have had to keep up. Countrywide, it has required wholesale improvements in almshouse accommodation – partly to compete with ‘new build’, and partly because of the demands of new legislation in areas such as building standards, health and safety, fire risk, and, not least, the maintenance of listed buildings in historic townscapes. They are no longer picturesque derelicts.

Burrill Homes, Water End, Clifton
Decorative Carving

10
Almshouses in an Architectural Context

Almshouse design

The architecture of almshouses can be very distinctive – they often form a line, or terrace, of small cottages under a common roof. The Merchant Taylors' *maison dieu* is a typical example, and Lady Hewley's is a slightly grander version. They were sometimes set round a court or garden, and those with large gardens cultivated them to provide vegetables, or kept chickens.

Courts were medieval institutions and included the royal court, parliament (itself a court), the courts of justice, the universities, and the monasteries. The constituent parts were literally built round a court, quadrangle or cloister and included the same ingredients – a domestic range, a place of religious worship and a business or meeting room – but with the emphasis on different aspects. They also might have, separately, accommodation for visitors, an infirmary for the sick, and lodgings for their elderly brethren or pensioners to pass their last years. These have all survived in their different ways, the latter in particular as almshouses. In the Netherlands where hundreds of old almshouses survive, one Dutch name for them is *hofje* or little court. This college-style arrangement of cottages round a quadrangle or court became common everywhere from the seventeenth century. In York, the twentieth-century Colton's Hospital, John Burrill Homes, John Hunt Homes and Ann Harrison's almshouses fall into this category, as do the Fothergil Homes in Avenue Road. And completely hidden behind its Ogleforth houses, Jane Wright's Saville Court is the latest example.

Those almshouses called hospitals tend to be large houses offering what became one- or two-room self-contained accommodation. Originally hospital inmates would have occupied beds lined up in a large hall or infirmary, usually with sight of a chapel altar. At some point when privacy became a social need, or when almspeople began to be residents rather than temporary inmates, partitions and walls would have divided the space into separate rooms. Examples of hospitals with private rooms built after the sixteenth century in York, are Ingram's, Anne Middleton's, Sir Thomas Hesketh's, the Old Maids' (Mary Wandesford's) and Dorothy Wilson's, all referred to as hospitals. In the twentieth-century the name hospital was often dropped in favour of "house", as at Wandesford House, or "home", as at the Walker Barstow Homes in Bishopthorpe Road. The courtyard almshouses also adopted this name.

Postwar demolition

The orgy of slum clearance and casual demolition of other buildings that took place in the 1950s and 60s in York, no doubt carried out for good reasons at the time, altered the appearance of parts of the city, and not necessarily for the better. Walmgate in particular, run down and with housing stock in very poor condition, suffered badly in architectural terms from the losses. The isolation of Bowes Morrell House in a wasteland of undistinguished 1960s flats is a monument to the kind of insensitive, or merely expedient, planning that prevailed in the postwar period. There was an understandable postwar taste for the new and clean in preference to the old and dirty – and much of it was very dirty as well as damp and badly decayed. Robert Winterscale's, built in 1723, was once part of that street, opposite St Margaret's church and Sir Robert Watter's Hospital is missing from the street scene off George Street, nearby. Other demolitions included the cottages of the Cordwainers on Piccadilly, Colton's in Tanner Row, and Barstow's off Nunnery Lane. Individually these may not have been of high architectural quality, though photographs suggest they were interesting, but they formed part of a distinctive and characteristic York scene among the brick terraces surrounding them.

Almshouses in a townscape

The intervention of John Bowes Morrell and Cuthbert Morrell, among other distinguished York conservationists, prevented even more losses and preserved many medieval, early modern and Georgian buildings, together with much of the charm of the centre of York. Some of that charm has arguably become a little fossilised to benefit the tourist industry, one of York's most important sources of income. Like Oxbridge colleges and cathedral closes, some of the more picturesque almshouse buildings form part of a public space, and as such, appear to be visitable tourist attractions. However, like colleges and canons' houses, they remain private domestic spaces where ordinary people live and whose privacy needs to be guarded. The small part that almshouses play in keeping an attractive street scene grounded in normal human behaviour also keeps the worst of the heritage industry at bay. Listed-building status too, though a burden in many ways, serves to maintain the architectural integrity of the older almshouses and of the street of which they form part.

York almshouses founded or rebuilt 1900-1983

The numbers refer to locations on maps 1-3 (p.69-71) and descriptions in Part II

(36) Ann Harrison's Almshouses 1845. Rebuilt 1964: Penley's Grove Street/Abbot Street.
(43) Lady Deramore's Memorial Almshouses 1903. The Village, Heslington
(44) Jackson Barstow's Almshouses 1910. Front Street, Acomb. Founded by J.J. Jackson Barstow. No longer almshouses.
(45) Colton's Hospital. Moved 1910 from Tanner Row **(28)** to Shipton St, Burton Stone Lane.
(46) Cameron Walker Homes 1912. Bishopthorpe Road: now **Walker Barstow Homes**.
(47) Burrill Homes 1931. Water End, Clifton.
(48) Fothergil Homes 1935–. Avenue Road, Clifton.
(49) Sir John Hunt Homes 1950/60s. Fulford.
(50) Jane Wright's 1983 (charity founded 1675). Ogleforth/Saville Court.

Earlier hospitals still functioning in the twentieth century

(7) Holy Trinity/Merchant Adventurers' Hospital c1371-3 Fossgate. Undercroft of the Hall. Closed in the early twentieth century.
(10) Merchant Taylors' *Maison dieu* rebuilt 1730. Aldwark, formerly the Confraternity of St John the Baptist. Became Caretaker's cottage in late 1940s.
(12) Cordwainers' *Maison dieu* 1811 in Church Lane, near St Denys's church off Walmgate. Originally built 1436. Demolished 1938.
(16) Sir Thomas Hesketh's Hospital 1608/1795. Heslington.
(17) Watter's Hospital 1609. Nowtegail Lane, (later George Street, off Walmgate). Rebuilt in 1844 in Chapel Row. Demolished 1958.
(19) Ingram's Hospital 1630-32, rebuilt 1649. Bootham. Closed 1957.
(20) Anne Middleton's 1659, 1771, 1828. Skeldergate. Closed 1970s.
(21) St Crux Almshouses 1663 probably in a court off Fossgate. 1895, Piccadilly. Demolished 1930s.
(24) Sir Henry Thompson's Hospital 1692. Castlegate. Demolished 1949.
(27) Dorothy Wilson's 1717, 1765, 1812. Foss Bridge. Sold 1999; plans to rebuild.
(29) Winterscale's Hospital 1723. Walmgate. Demolished 1957.
(30) Wandesford's Hospital 1739. 37 Bootham.
(31) Margaret Mason's 1732. 21, Colliergate. Closed 1958.
(33) St Martin's cottages pre-1824. St Martin's churchyard, Coney Street, sold 1949. Mostly demolished.
(34) St Catherine's 1835. Holgate Road. Originally The Mount **(6)**. Demolished c1962.

(35) Lady Hewley's 1840, St Saviourgate. Originally Tanner Row **(26)**.
(36) Ann Harrison's 1845. Penley's Grove and St John Streets. Rebuilt 1964.
(37) Bishopthorpe Poorhouse 1846. Main Street, Bishopthorpe. Demolished 1963.
(38) Barstow's Hospital 1859. The corner of Caroline Street and Clement Street at the Clementhorpe end of Nunnery Lane. Originally Blossom Street **(25)**. Demolished 1960s.
(39) St Thomas's Hospital 1863. Victoria Bar, Nunnery Lane. Originally at Micklegate Bar **(8)**. Closed 1972.
(40) Sister Wilson's 1885. East Parade, Heworth. Demolished c1957.
(41) Ellen Wilson's 1894. Lawrence Street, next to St Lawrence's church.
(42) Terry Memorial Homes 1899. Skeldergate, in front of Anne Middleton's. Closed 1970s.

Almshouses still functioning in 2010

(16) Sir Thomas Hesketh's 1608/1795. Heslington.
(30) Wandesford House founded 1725, built 1739. 37 Bootham.
(35) Lady Hewley's 1840, St Saviourgate. Originally 1700, Tanner Row **(26)**.
(36) Ann Harrison's 1964, Penley's Grove Street. (Original foundation of 1845 demolished).
(41) Ellen Wilson's 1894. Lawrence Street, next to St Lawrence's church. May merge with **Dorothy Wilson's (28)**.
(43) Lady Deramore's Memorial Almshouses 1903. The Village, Heslington.
(45) Colton's. Moved 1910 from Tanner Row **(29)** to Shipton Street.
(46) Walker Barstow Homes. 1912. Bishopthorpe Road.
(47) Burrill Homes, 1931, Water End, Clifton.
(48) Fothergil Homes, 1935, Avenue Road, Clifton. Enlarged several times since 1957.
(49) Sir John Hunt Homes, 1950/60s, Fulford.
(50) Jane Wright's, 1983 (charity founded 1675), Ogleforth/Saville Court.

11
Almshouse Regulation

Almshouses set up through benefactions were governed by the original wishes of the benefactor. Criteria for the selection of beneficiaries, for example, were often included in wills. The most frequently cited criteria were simply that beneficiaries should be in hardship, distress or need, and be over the age of about fifty or sixty. Sometimes places were available only to people from the immediate area, or who had some link with it. Often there was a religious requirement, whether to be a member of the established church, as at Mary Wandesford's, or Unitarians as at Lady Hewley's. Neither almshouse insists upon these criteria now, though applicants are expected to be practising Christians. There were a great many which took in only women - spinsters without family, and widows with no other means of support were among the poorest of the deserving poor.

In the UK, about 36,000 older people live in 2,600 groups of almshouses under the control of about 1800 separate almshouse charities. Most are very small, housing around eight to twelve people each; a few are very large. Many of these larger almhouse charities are housing associations. One such is in Hull: Pickering and Ferens Homes, an amalgamation of two almshouse charities which, passing from council ownership, became a housing association now owning nearly 1200 properties. The largest almshouse in York, Fothergil Homes (which is not a housing association), houses some sixty people.

Charity Commission

To register with the Charity Commission, an organisation must demonstrate a charitable purpose. Most almshouse charities register with the Commission because their aims are, among other things, the prevention or alleviation of poverty and, in particular, the relief of those in need by reason of their age. Registration requires a governing document, a Charity Commission scheme, to be drawn up, and schemes are usually based on the original selection criteria. Until fairly recently the Commission held out strongly against modification of criteria which were based on the original intentions of the benefactors. It was only in 2000, for example, that Mary Wandesford's charity scheme was redrawn to permit widows and members of any Christian denomination to apply, and it took until 2007 for the charity to be permitted to accept divorced women. Registration with the Commission makes all the activities of a charity transparent. The annual financial statements of larger charities, together with details of trustees and clerks, as well as the charity's aims can be readily examined by anyone with access to the Internet.

Almshouse Association

The Almshouse Association, of which many almshouses are members, exists to protect their interests. Almshouse charities pay a subscription to fund the aims of the Association, which are to maintain standards, to provide training courses for clerks, wardens or scheme managers, and trustees, and to give advice. It publishes a guide to almshouse management and a specimen residents' handbook. It also engages to collect evidence of problems affecting almshouse charities caused by government legislation, and make representations to the relevant government department (for example, the Department for Communities and Local Government).

Unfortunately, it is not necessarily consulted by government in advance of legislation that might affect the sector. There have been changes, for example, in how Council Tax is charged which affects almshouses. The effect on charities was not considered by government before the legislation was changed to enable councils to make a ninety percent charge for second homes and empty properties. Many councils make no distinction between dwellings left deliberately vacant, and charity property (particularly bedsit flats) not let because occupants cannot be found. The first six months of almshouse vacancy are zero-rated, but thereafter very few authorities charge just the minimum fifty percent set by the government. York is not one of these noble few and the result is a ninety percent tax on charity's lost income.

The question of registration under the Safeguarding Vulnerable Groups Act 2006 has also been examined and commented on by the Association in order to provide guidance to its members. A possible result is that a new government is now (2010) re-examining the legislation.

The Housing Corporation and Registered Social Landlords

The Housing Corporation was set up as a government agency under the Housing Act 1964 and has performed a role in central government's social housing policy and regulation ever since.

In 2004 the Housing Corporation reported in no. 41 of its online report series, *Sector study*, that about a quarter of almshouses, some 416, were registered as social landlords (RSLs) with the Housing Corporation. This represents about one percent of the total housing stock held by registered social landlords. Few almshouse charities wish to register, however.

Registration enables an almshouse to apply for funding for the development of new homes, or for major repairs and development; it establishes credibility and status to borrow money from other sources; and it ensures that residents are treated appropriately under the Housing Corporation's regulatory code. However, that code is somewhat onerous and RSLs are not necessarily as free about choice of residents. This is true, too, for almshouses which are housing associations. These are an overlapping group, many members of which are also RSLs and charities.

Housing Associations

Housing associations are often seen as a development of medieval and subsequent models of social housing provision. This is not necessarily the case; they have actually emerged more from self-help groups than from the top-down aid to the poor given as benefactions. They are not-for-profit bodies providing low-cost social housing for people in housing need. They are also more associated with the political relationship between central and local government. When local government, and state provision generally, was strong after 1945 they were barely tolerated. Since 1979 government has become more centralised and as local government has increasingly lost power over local affairs housing associations have increasingly become instruments of central government's housing policy. One of the areas local authorities lost control of was provision of social housing, and after those changes were brought in by the Thatcher government housing associations became a more important player in the housing sector. Local authorities sold off their council housing stock or transferred it to housing associations which were able to apply for funding from housing benefit schemes not available to the local authorities. Many former municipal almshouses became housing associations in order to take advantage of those loans and grants and upgrade the buildings to a higher standard; and some have become very large. There is now a large quantity of purpose-built accommodation for vulnerable people, including the elderly poor and infirm.

Housing associations were regulated and funded until 2008 by the Housing Corporation, but now fall under the aegis of two separate bodies dealing with funding and tenants. Now that they are so closely associated with government policy and funding, they are at risk of losing their independence and influence.

Trustees and clerks of almshouses

Almshouse charities are governed by trustees, volunteers who must not gain any financial benefit, or income, from the post. There is an increasing need to find and appoint trustees with relevant experience and expertise, who also have spare time. The day-to-day running of an almshouse is usually carried out by a clerk or scheme manager, sometimes by a full-time caretaker or warden. Most have both clerk and caretaker. These posts are usually paid, unless the work is carried out by a trustee. Increasing legal requirements placed on small businesses and charities make the employment of professionally qualified staff significantly desirable, and in York many almshouses are run from the offices of solicitors or estate agents, businesses with (in theory) appropriate qualifications and experience already in-house. Desirable areas of expertise include tax (especially PAYE and VAT), accounting, building regulations, fire risk analysis, business and staff management – as well as awareness of the potential care needs of the elderly.

When appointing new staff, almshouses are encouraged to have Criminal Records Bureau checks carried out – these have to be done by umbrella organisations because of the expense. There is an additional possibility, mentioned above, that all staff, all trustees, and all contractors working for almshouses may have to register under the Safeguarding Vulnerable Groups Act 2006. Almshouses which do not qualify in any way as sheltered housing may escape this requirement. The definitions have not been tested in the courts but as it comes at a financial cost, it may have a profound effect on the willingness of the newly retired to offer their services as clerks and trustees. Whether this (apparently one-off) registration, or the CRB checks, serve their intended purpose is open to doubt. It may be that a government with a less-controlling disposition will reconsider the question.

Residents of almshouses

Residents of almshouses are appointed as charity beneficiaries, as they were once appointed to endowed places or beds. They are selected through their qualifications under the terms of the charitable trust, often involving examination of their financial resources, and at the discretion of the trustees. They gain no property rights by accepting an offer of accommodation because charity property is inalienable. They do not become tenants, instead they become licensees, permitted to live in charity property under the terms of a contract. They pay weekly maintenance contributions, rather than rent, though these are normally based on rents in the area (and often set lower). The range of amounts charged is wide. The Housing Corporation's *Sector study* 41 gives the national average in 2004 as about £52 (from a range of zero-£146), and the average for Yorkshire and the Humber as about £56 per week (from a range of about £14-£96). These figures were lower than the figures for average regional rents for other social landlords, but some almshouse residents also pay an additional sum for heating and lighting, costs which are not eligible for Housing Benefit.

Residents need to remain healthy and independent or they may be asked to move into residential care or the care of a hospital (medical or mental). The provision of social care by the state has made it possible for almshouses to help residents remain in their homes much longer, and often for life, without needing to provide what would be very expensive care themselves.

The qualifications above suggest a somewhat restrictive regime. In practice, almshouse residents live in pleasant, comfortable and often companionable circumstances. Private charity remains for the most part long-suffering and kind.

St Anthony's Hall, Peasholme Green

Part II

Checklist of named almshouses and hospitals founded in York, 936-1983

The locations of many lost almshouses are shown on the 1852 map of York, a very large-scale plan of the city carried out by Captain Tucker R.E. for the Ordnance Map Office. Some of the medieval sites, it should be noted, are misnamed or incorrectly dated. The map is held in the York City Archives, and it has been made available online at http://york1852.org/ through a collaboration between Beacon Dodsworth and the City of York Council. Almshouses marked on the 1852 Ordnance map are indicated below with a hash #.

Medieval almshouses in York

The numbers refer to locations on maps 1 and 3 (p.69 & 71), and to the descriptions following in this section.

(1) St Leonard's# c936, originally St Peter's. SW corner of the Roman fortress to Bootham Bar. (Some of its ruins survive).
(2) St Nicholas's# 12th century. Near 148 Lawrence Street.
(3) St Giles without Walmgate 1274. Lawrence Street.
(4) St Loy# / St Leonard 12th(?) century. Monk Bridge
(5) St Mary in the Horsefair# 1315/18. Horsefair (Union Terrace/ Clarence Street).
(6) St Catherine's 1333. The Mount, near 116-128 Blossom Street. See also (**34**).
(7) Holy Trinity# c1371-3. Fossgate, the undercroft of the Merchant Adventurers' hall. (The building survives).
(8) St Thomas's# 1390. Next to Micklegate Bar. see also (**39**).
(9) St Helen's fl. 1399-1662. Fishergate.
(10) Confraternity of St John the Baptist# *Maison Dieu* c1415. Aldwark. Rebuilt 1730 for the Merchant Taylors guild. (The building survives).
(11) St Anthony's, Gillygate founded before 1420. Horsefair (Junction of Wigginton Road and Haxby Road).
(12) The Cordwainers' *Maison dieu*# 1436. Rebuilt 1811 in Church Lane near St Denys's church, off Walmgate (now Piccadilly).
(13) St Anthony's# 1446. Peasholme Green. (The building survives).
(14) Common Hall *Maison dieu* and chapel# mid-fifteenth century. Coney Street/Lendal.
(15) St Mary Magdalene# before 1481. Clifton, near the city boundary, Burton Stone Lane.

Post-Reformation almshouses in and near York

The numbers refer to locations on maps 2 and 3 (p.70-71), and to the descriptions following in this section.

(16) Sir Thomas Hesketh's Hospital. 1608/1795. Main Street, Heslington. (The almshouse continues).
(17) Watter's Hospital# 1609. Nowtegail Lane, (later George Street, off Walmgate). 1844, rebuilt in Chapel Row#
(18) Agar's Hospital# 1631 Monkgate.
(19) Ingram's Hospital# 1630-32, rebuilt 1649. 90 Bootham. (The building survives).
(20) Anne Middleton's Hospital# 1659. Rebuilt 1771 and 1828. Skeldergate. (The building survives).
(21) St Crux Almshouse 1663 probably in a court off Fossgate. 1895, Piccadilly.
(22) Ann Wright's Hospital 1670. Bishophill. Location unknown.
(23) Poor's Cottages 1691. Front Street, Acomb. Demolished 1950.
(24) Sir Henry Thompson's Hospital# 1692. Castlegate.
(25) Barstow's Hospital# early eighteenth century. Blossom Street, near Holgate Road turn. See also (**38**).
(26) Lady Hewley's Hospital 1700. Tanner Row. See also (**35**).
(27) Dorothy Wilson's Hospital# 1717, rebuilt 1765 and 1812. Foss Bridge. Sold 1999. (Plans to rebuild).
(28) Colton's Hospital# 1717. Tanner Row. See also (**45**)
(29) Winterscale's Hospital# 1723. Walmgate.
(30) Wandesford's Hospital# founded 1725. Built 1739. 37 Bootham. (The almshouse continues).
(31) Margaret Mason's Hospital# 1732. 21 Colliergate.
(32) Briggs' Charity 1802. King Street; moved to Low Ousegate 1860s; defunct before 1907.
(33) St Martin's Cottages pre-1824. St Martin's churchyard, Coney Street, sold 1949.
(34) St Catherine's Hospital# 1835. Holgate Lane (later 45 Holgate Road). See also (**6**).
(35) Lady Hewley's # 1840. St Saviourgate. See also (**26**). (The almshouse continues).
(36) Ann Harrison's Almshouses# 1845. Penley's Grove Street/St John St. Rebuilt Abbot Street, 1964. (The almshouse continues).
(37) Bishopthorpe Poorhouse 1846. Main Street, Bishopthorpe.
(38) Barstow's Hospital 1859. Caroline Street/ Clement Street, off Nunnery Lane. See also (**25**) and (**46**).
(39) St Thomas's Hospital 1863. Victoria Bar, Nunnery Lane. Closed 1972. See also (**8**). (The building survives).
(40) Sister Wilson's Hospital 1885. 17-19 East Parade, Heworth.
(41) Ellen Wilson's Hospital 1894. Lawrence Street, next to St Lawrence's church. (The almshouse continues).
(42) Terry Memorial Homes 1899. Skeldergate, in front of Anne Middleton's. Closed 1970s. (The building survives).

Twentieth-century almshouses in and near York

The numbers refer to locations on maps 2 and 3 (p.70-71), and to the descriptions following in this section.

(43) Lady Deramore's Memorial Almshouses 1903. The Village, Heslington. (The almshouse continues).
(44) Jackson Barstow's Almshouses 1910. Front Street, Acomb. (The building survives).
(45) Colton's Hospital 1910. Shipton St, off Burton Stone Lane. See also (**29**) (The almshouse continues).
(46) Cameron Walker Homes 1912. Bishopthorpe Road: now **Walker Barstow Homes**. (The almshouse continues).
(47) Burrill Homes 1931. Water End, Clifton. (The almshouse continues).
(48) Fothergil Homes, 1935–. Avenue Road, Clifton. (The almshouse continues).
(49) Sir John Hunt Homes, 1950/60s, Fulford. (The almshouse continues).
(50) Jane Wright's Almshouses 1983 (charity founded 1675) 3-9 Ogleforth/1-5 Saville Court. (The almshouse continues).

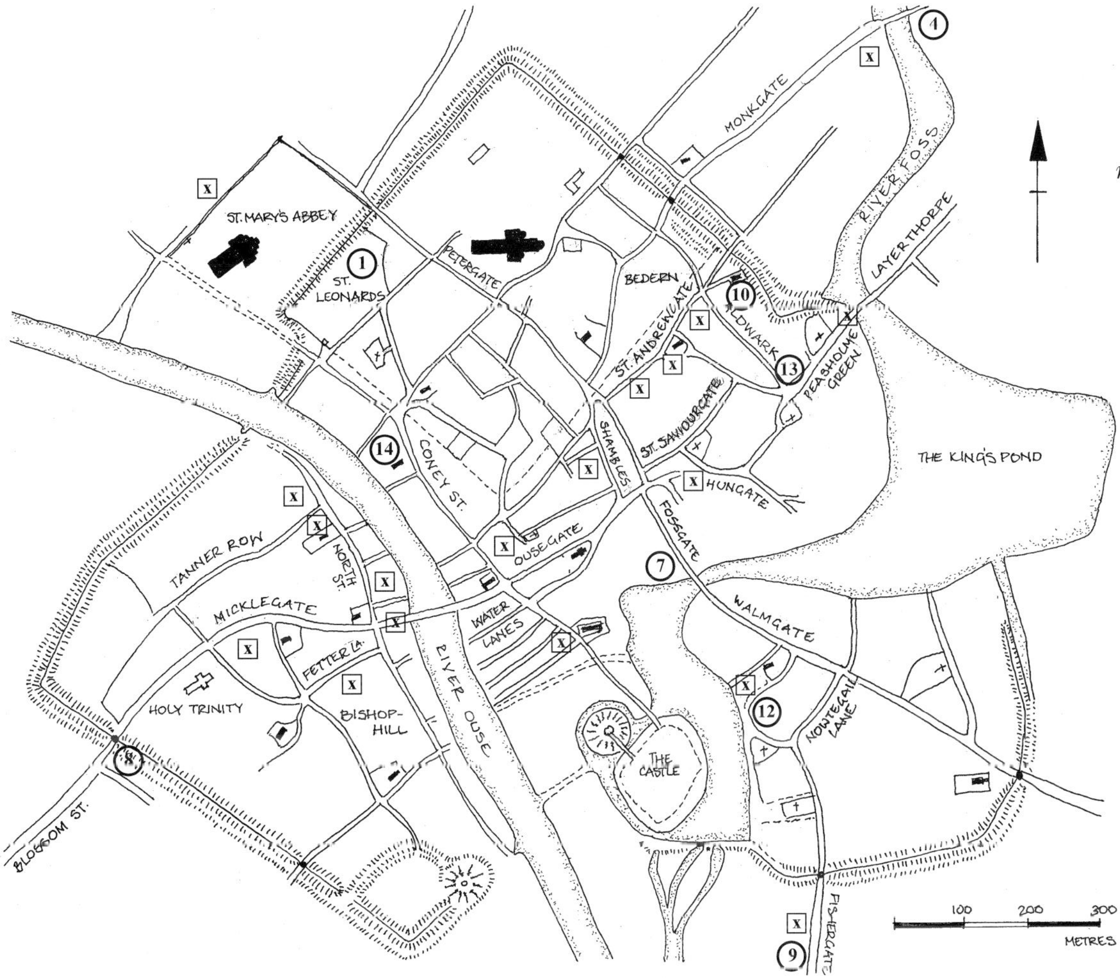

Map 1
Medieval York

Central hospitals and almshouses, nos 1-15, *maisons dieu* marked [x] (see also Map 3)

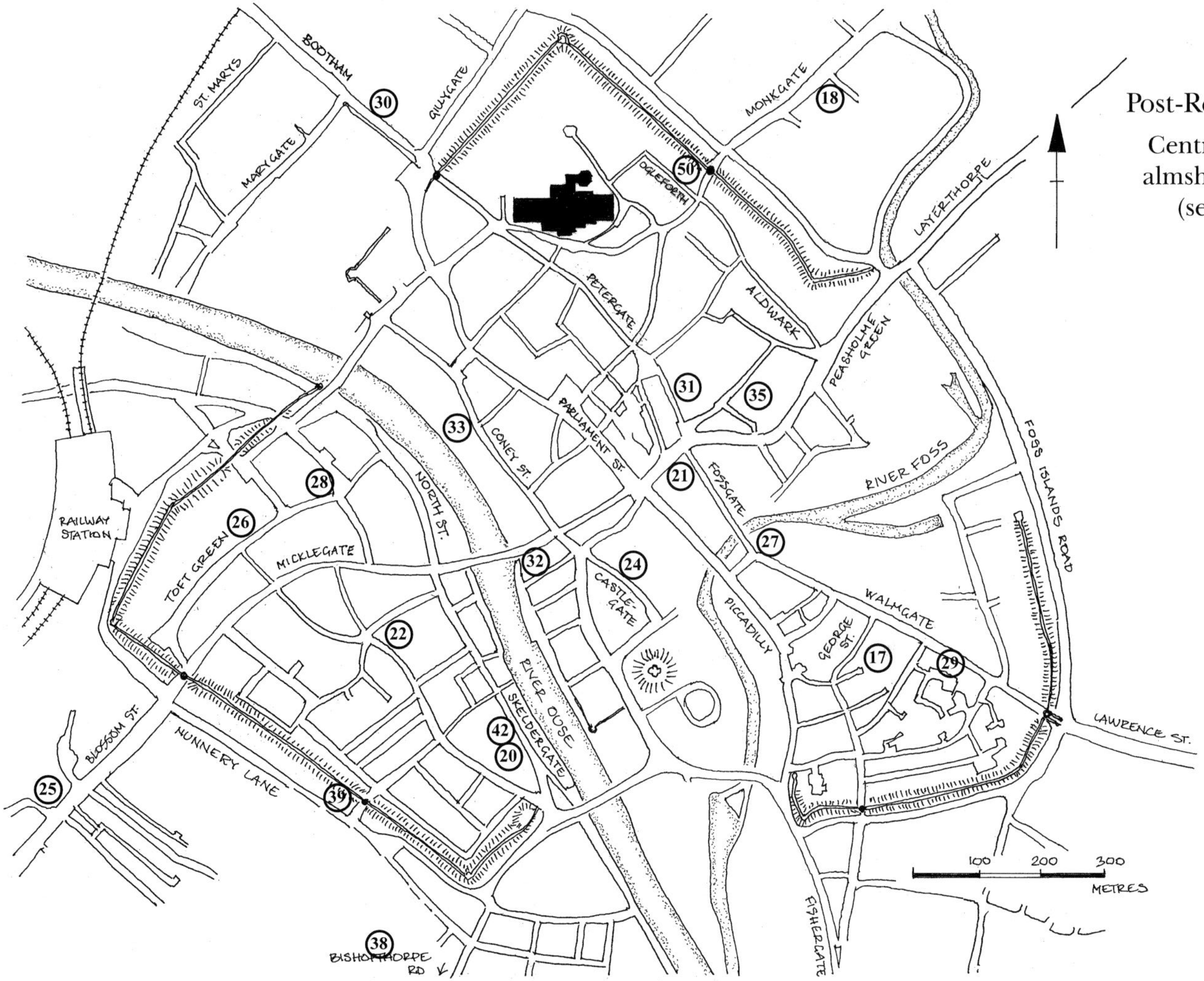

Map 2
Post-Reformation York.

Central hospitals and almshouses, nos 18-50 (see also Map 3)

Map 3
Outer York

Hospitals and almshouses, medieval to modern, nos 2-49

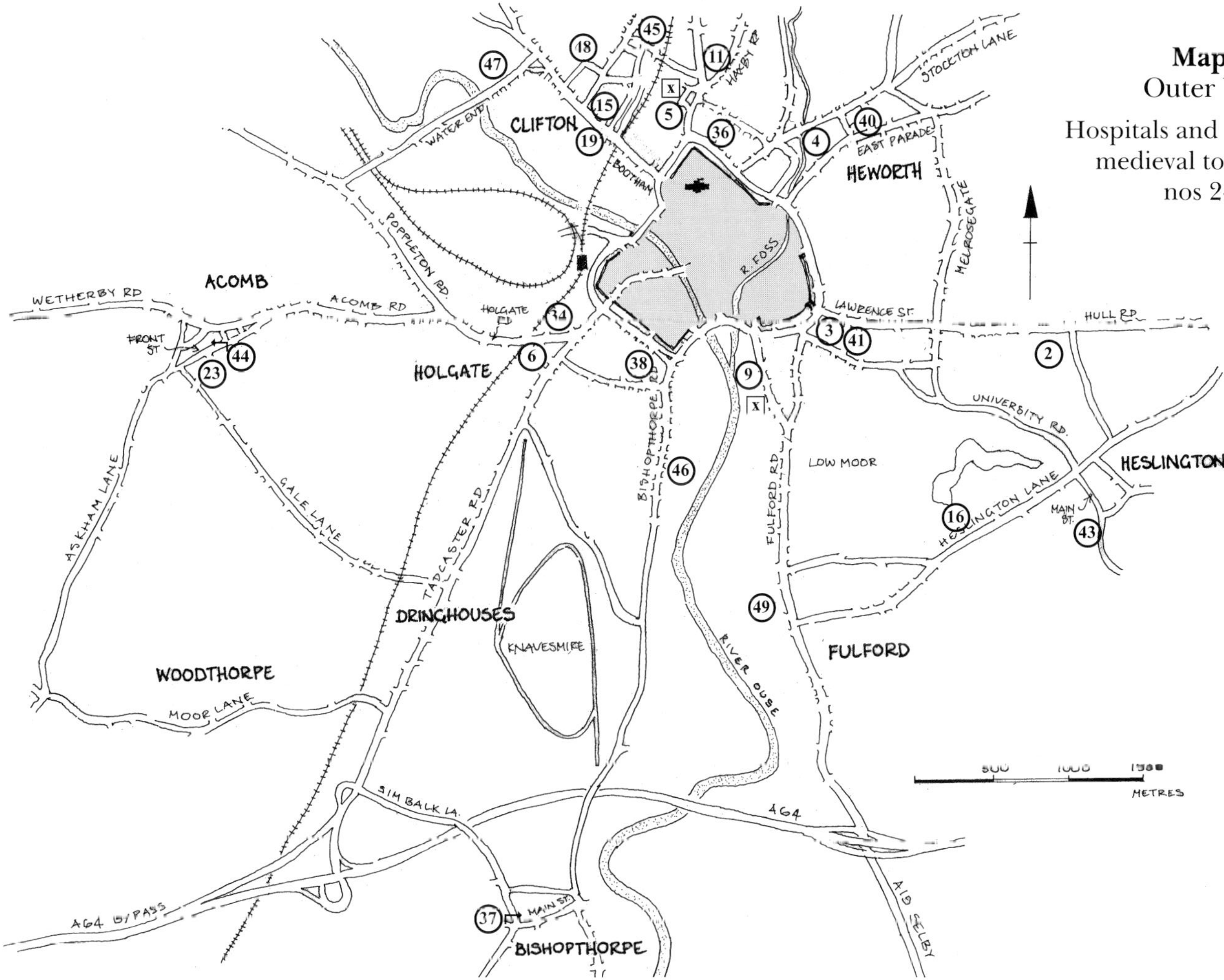

York Hospitals and Almshouses: Descriptions

The descriptions below are roughly in order of the date of original foundation; the numbers refer to the map references for each establishment. Details have been gathered from a range of published sources and, as records for some are very limited, those descriptions are correspondingly short. Included in the descriptions are some of Bill Sessions' unpublished manuscript notes from visits he made to still-functioning almshouses in 1946.

(1) St Leonard's Hospital

The legend of King Athelstan's gift

According to legend, King Athelstan passed through York in about 936 after his victory over the Viking armies at the battle of Brunanburh and observed the *culdees*, or secular canons of the cathedral, providing alms to the poor from their own slender resources. His gift of money and an endowment, enabled the construction of a dedicated almshouse or hospital on royal land close by, somewhere to the west of the Minster, to accommodate some of the very poor, elderly and infirm people of the city, some of whom were sleeping in the streets. It was dedicated to St Peter, and may have been the very first purpose-built almshouse in England.

That it was a pre-Conquest foundation seems likely, but it is not clear whether the story of Athelstan's gift is true, so it is worth examining its origins. The story has been repeated in histories of almshouses many times, from William Dugdale's *Monasticon anglicanum* to T.M. Fallow who researched the story for his essay in VCH (v. 3, 1913), but the fullest and best account is P.H. Cullum's in her doctoral thesis on charities and hospitals in medieval Yorkshire (1989). She doubts the Athelstan story, but inclines to the Anglo-Saxon origins of the hospital for reasons partly to do with two words of pre-Conquest origin, *culdees* and *thraves,* which continued to appear in the post-Conquest records. She believes the almshouse was in existence before Athelstan's endowment because of its original link to the Minster – which, if true, would make it even older and therefore certainly the first almshouse in England.

The secular canons at St Peter's cathedral church in York (and at Beverley) are referred to in the records of the Hospital as *culdees, colidei, kolidenses* and other variants of the gaelic word *cele-de*. The name derives from the Celtic church and means servant of God. They were originally an ascetic sect attached to monastic foundations who gave alms, or doles of (probably) food, to the poor from an "alms house", originally perhaps some sort of office or chapel. The word *culdee* is, significantly, not found in Minster records, only in those of the Hospital.

The 1173 cartulary of the hospital in the British Library (Cotton Ms. Nero D. iii) contains an *Historia fundationis*, the story of its pre-Conquest foundation. It describes Athelstan's visit and his endowment of a thrave from every plough in the extensive diocese of York. *Thrave* is a Scandinavian word for a measure of grain: either twelve or, more usually, twenty-four sheaves of corn (two stooks), which were originally royal dues given to holy places. The diocese included Lancaster, Westmorland and Cumberland, so the endowment, later known as Petercorn, was generous. The *Historia* was evidently a compilation from various chronicles, all keen to link the hospital with the famous King, which seems to indicate that the idea of its ancient origins was already established by the eleventh and twelfth centuries. There is a line in the charter which states that the endowment was said to be the reason for the hospital's existence: *illam antiquam elemosinam supra qua dictum hospitale fundatum existit*. Fallow (VCH 1974/1913) quotes from an inquiry reported in Assize Rolls 1045, m.17d that "at the time of the Conqueror after an ancient war the clerks of the church of St Peter ... asked the King to give them a place lying before the gate of the church on the west as a site for buildings to receive and lodge the poor sick and infirm who were at that time suffering from extreme want, lying by night in the streets".

It seems that the endowment was withdrawn soon afterwards – regranted to the Crown for its original purpose of exterminating wolves. Though this sounds improbable, wolves were a problem. A wayfarers' hospital at Flixton, Holderness existed "to preserve travellers from being devoured by wolves and other voracious beasts" (Clay, 1909) The thraves were reinstated by William I in 1069 at the request of the Archbishop. If this is true his eleemosynary motives may have been mixed. Perhaps it was remorse (or blood money) for the recent Harrying of the North, and burning of the city – but a good deal of the land supporting the endowment must have been devastated too, and therefore unable to pay, at least in the short term

The serious fire in York in 1069 (part of the Harrying of the North), and another traditionally in 1137, are both thought to have caused the hospital to move its position. Professor Christopher Norton convincingly argues that the second *conflagrata* may be a misreading of *consecrata* in a document mentioning it – not a major fire but a consecration of churches. There is no archaeological evidence for a second fire, which in any case would have had to affect all the named churches listed in the document on both sides of the river. The hospital was certainly destroyed in the fire of 1069, along with the Minster and much of the city, which may have occasioned its move into the south-west corner of the Roman fortress, where some of its ruins remain to this day. William II (William Rufus, son of the Conqueror), who confirmed the thraves as ancient alms, is said to have enabled its move into this position. It would anyway have had to move away from the Minster to enable the cathedral's rebuilding and expansion in 1079. Whether there was another interim move is impossible to say.

A document in the Patent Rolls of Edward III, 1334-8 (pp. 266-8) confusingly suggests that William II was the founder of the hospital, built the chapel of St Peter and endowed it with the thraves. King Stephen, grandson of the Conqueror, was also claimed as the second founder (after Athelstan). He later had a chapel constructed within the hospital, or nearby, in honour of St Leonard and from the thirteenth century the hospital became known as St Leonard's, though it retained its original seal with the figure and name of St Peter.

St Leonard's Hospital.
Ruins of the infirmary

The hospital

Whatever the truth of its origins, this famous York hospital became possibly the largest in England and, in York, an institution third only in value after the Minster and St Mary's Abbey. It was capable of housing more than 200 sick poor people, not all of whom were chronically sick – this was also a hospital for acute cases, and tried to cure the sick not just house them. At least one of the sisters looking after the sick appears in the records as *medica*, a doctor. The religious brothers, however, were subject to the Rule of the religious foundation and studied, worked and prayed accordingly.

Some of the inmates were known as "cremetts" – a local dialectal word (originally from eremit/hermit) meaning weak and infirm - who would have been housed and treated free. There was also a group of boarders known as corrodians. A corrody was bought with cash or a donation of land or property, and enabled the corrodian to live for the rest of his or her life in the hospital. The hospital also provided outdoor relief in the form of support for some thirty regular dependants, livery-holders and pensioners, as well as beggars at its gates.

There were a number of orphans, sometimes more than twenty, cared for by eight Augustinian sisters and some lay brothers. The orphan children were housed in the "barnhous" (another word of Scandinavian origin – bairn is still northern-English usage for child). There were grammar and singing masters teaching the choristers, and though there was originally a distinction between orphans and choristers, this had apparently disappeared by the sixteenth century.

St Leonard's Hospital.
Ruins of the infirmary undercroft inside the Roman fortress.

At around four acres, the hospital was the size of a small village, and was independent of the Minster, the city, and the guilds. It was also responsible for supporting and supplying food to other hospitals, particularly the leper hospitals and also the prison, so all related services would have been there. As an establishment with the rights of manorial courts, it even had its own gallows (outside the city on Garrow Hill). Unusually, if not uniquely for a hospital, it also established daughter hospitals on its lands, at Bowes, Hedon and possibly at Tickhill. There was also a leper hospital at Newark under the direction of Robert the Almoner of St Leonard's.

The endowment of thraves held its value, being worth more than twice the income from the manors in the fourteenth century. Payments of them had to be protected because those paying found it onerous. However, various corrupt masters in the fourteenth century embezzled the income and caused the hospital to decline. In 1469 the payment of the endowment was one of the causes of an East Riding uprising, when 15,000 men marched (unsuccessfully) on York. As a result the thraves were stopped by Edward IV without compensation, which led to a further decline in the hospital's already sinking fortunes. The numbers of poor it was able to support dropped from 130 to about sixty.

By the sixteenth century its income had dropped to a third of what it had been at its height in 1280. Its continuing value as an institution, however, was recognised by the authorities, who probably did not intend its dissolution. Nevertheless its master Thomas Magnus surrendered it, and by 1540 it had been dissolved. This has been construed as a disaster for the hospital inmates, and for the sick in York generally, particularly if the King had intended its continuation under secular authority.

The evicted poor and sick may have continued to receive "Cremett Money" – payments appeared under this heading (spelt "Cremitt" later) thereafter in the city accounts. There was evidently no money in the city accounts to purchase the hospital for continued use. This may have been partly due to a desire not to provoke the King, as well as to the fact that the structural deterioration would have been expensive to put right.

The hospital building was probably in one of two courtyards: the other contained the buildings relating to the religious foundation. The whole site was bounded by walls, and occupied all the land inside the city walls between the Roman fortress wall and Bootham Bar. This very large area includes what lies under the present St Leonard's Place and the Theatre Royal, back to the rear of High Petergate. A lane ran alongside the boundary wall down to the river, which was called Footless Lane at one time, and Finkle Street later – names which may refer to the crowds of bent cripples waiting at the hospital gates, or possibly to a kink in the lane. Widened, it is now Museum Street, named for the Yorkshire Museum in the grounds of St Mary's Abbey. The gatehouse of the hospital is said to lie under the Red House opposite the end of Blake Street, with some of the original boundary wall forming part of the Theatre Royal's offices next door. Hargrove (1818) refers sadly to the demolition of the walls in 1782, carried out to enable carriages to get to the Theatre.

Of the original buildings, there remain only parts of the infirmary (or chapel), and its undercroft, next to the Roman fortress walls overlooking the Museum Gardens. That part of the hospital was used as the Royal Mint from the late sixteenth century, and the area around it, once known as Mint Yard, is now occupied by the city library. A lane led through to other parts of the Mint, where the Theatre Royal now stands, until the construction of St Leonard's Place in the 1830s.

The name was transferred in 1985 to St Leonard's Hospice on Tadcaster Road which continues the tradition of care.

(2) **St Nicholas's Hospital** Walmgate Without (Lawrence Street)

The largest leper hospital in York was St Nicholas's outside Walmgate Bar, located to the east of St Lawrence's church in the vicinity of no. 148 Lawrence Street and under the Bootham Engineering site. It is also marked on a map in Francis Drake's *Eboracum*. Reminders of its existence are retained in place names such as St Nicholas Gardens and St Nicholas Fields. It was a well-endowed institution which may account for its receiving very few individual benefactions – it appeared to benefactors not to need them.

The church may have been founded between 1088 and 1112 on

land given by Stephen, the first Abbot of St Mary's, with the hospital founded a little later, as grants of land were also made in 1132 and 1161. The Empress Maud (or more probably her mother, Edith Matilda, Henry I's queen, who devoted much charity to lepers) gave lands for its support on condition that, on the eve of Saints Peter and Paul, bread, ale, cod or salmon, butter and cheese should be given to the lepers.

In the fourteenth century there were reports of abuses and disobedience on the part of the religious members of the community; it declined and was later annexed by Holy Trinity Priory. That period seems to have been a bad time generally for these institutions. It also suffered from having a master who appropriated hospital property for himself at the end of the century. He was John Midelton, who was also Richard II's doctor, and presumably therefore an absentee master. The hospital declined and may subsequently have become a house of retirement for women not accepted for the Clementhorpe nunnery.

Parts of the site were excavated in 1992 and 1993 and revealed that the main building was an aisled structure with a large hearth. At its north end the building was divided into cubicles with their own hearths – where the lepers, or perhaps the brethren, may have lived. The hospital was also surrounded by substantial boundary ditches.

The hospital's church acted as a parish church, and survived the Reformation but was destroyed in 1644, during the Civil War siege. As a sniper post it was bombarded from Lamel Hill above. Masonry from the hospital may have been used to repair Walmgate Bar in 1648, and to provide paving in Lawrence Street in the eighteenth century. Doorways from the church are said to have gone to both St Margaret's and St Denys's in Walmgate. The bells went to St John's church at the bottom of Micklegate, currently occupied by "The Parish Lounge Bar", where they hang above the heads of the bartenders, and are rung from time to time by members of the local association of bellringers.

(3) St Giles without Walmgate

There is very little known about this hospital. It was probably founded in the late thirteenth century, around 1274. It may have been a chapel of St Leonard's Hospital; it was certainly too poor to be included in valuations at the end of the century. Like the hospital of St Giles in Beverley (a much larger establishment), it is known to have had an association with housing an anchorite and recluses, who would have attracted almsgiving or gifts. Its association with the great hospital of St Leonard's would otherwise have made such giving seem unnecessary.

(4) Hospital of St Leonard (or) St Loy
Monk Bridge

St Leonard's leper hospital may have been founded in the twelfth century. It may be a separate foundation from St Loy; it may have superseded it or have been renamed St Loy later. St Loy is an anglicisation of St Eligius or Eloi, sometimes invoked as the patron saint of the poor and poorhouses. Chaucer's Prioress, Madame Eglantine's "greatest oath was but by St Loy" – a rather mild sort of oath, suitable for a nun.

The hospital is said to have been on the east side of Monk Bridge. Its site is marked on the 1852 map as the "supposed site of the hospital of St Loy". It may have been in some way related to Robert Holme's *maison dieu* at that end of Monkgate or, indeed, a re-foundation by him. If so, it would have received the benefit of his bequest in 1396 that his relatives support his twenty-bed hospital for a hundred years after his death. The hospital had been granted protection by Edward III in 1350 to enable it to be supported by alms. This was a licence to beg, which was otherwise illegal. The gatehouse was apparently incorporated in a later building which was demolished in 1925, and nothing above ground now remains, though excavations in a car park on the site in 2001 revealed stonework which may have been part of it.

(5) St Mary in the Horsefair

This hospital, dedicated to the Virgin, was established outside the city walls on the site of the former Carmelite Friary in the Horsefair (near the junction of Union Terrace and Clarence Street). The friary was founded in 1253 but had moved by 1300 to its later site between Fossgate and Hungate. The hospital which took over the buildings originated with a chantry chapel in 1315, and was enlarged into a hospital in 1318 for a master and six infirm and aged chaplains. It is variously referred to in the records as "St Mary Bootham", and "Our Lady in Bootham, called the Horsefair". It was founded by Dean Robert Pickering, an exceedingly wealthy cleric. In 1556 it was leased to John Dawson of Heworth for a rent of 26s 8d but, apparently empty and ruinous in 1557, it was annexed by the Dean and Chapter and granted to a former priest of the hospital. It quickly became, or was taken over by, the grammar school of St Peter to house fifty boys and their teacher. They had previously been housed in St Mary's Abbey and, when that was dissolved, needed accommodation. The school was forced out by the Civil War siege of 1644 when the building was burnt down. The school either moved back into its remains, or simply retained the name "the Horsefair school", for some years afterwards until it transferred to the College of the Vicars Choral in Bedern.

William Hargrove (1818) refers to this establishment as the Hospital of St Mary Magdalene founded in the Horsefair in 1330 for aged priests by Robert de Pickeryng, Dean of York. It is marked with this date on the 1852 map of York on the site of the 1315/1318 foundation. Hargrove clearly means the same hospital as he also says that the lands of the hospital were devoted to the establishment of the free grammar school (which in his time was in St Andrew's church, St Andrewgate). The confusion over the date may be due to the date of Robert Pickering's death in 1332, but the confusion over the dedication to the Magdalene is less clear. "Our Lady" and "the Magdalene" were not, of course, the same person. Probably Hargrove confused it with the hospital in Clifton **(15)** below.

The site was excavated in 1972 when much of Union Terrace was demolished. It is now, appropriately enough, the location of the (somewhat unpopular) Arc Light hostel for the homeless.

(6) St Catherine's Hospital Blossom Street (34) Holgate Road.

St Catherine's Hospital was founded in the early fourteenth century as a leper hospital. It therefore stood well outside the city walls, probably near nos. 110/110a Blossom Street (currently Kwikfit garage), or a little further out in the area of nos. 116-128. In 1333 Edward III granted it protection for two years, enabling leprous men of the hospital to collect alms. It actually housed both sexes, and having become a city charity it escaped suppression in the sixteenth century. It consisted of a small group of buildings surrounded by a wall, with separate accommodation for men and women. With the decline of leprosy, it became a hospital for the elderly and infirm.

It was in a ruinous state by 1648, possibly as a result of damage caused during the siege in 1644, though that area was successfully defended by the Royalist forces which held both Micklegate Bar and the earthwork fort or sconce on the Mount behind the hospital. It is not known how the poor people living in it were protected, or where they went during the siege, but it was taken down and rebuilt in brick in 1652. The number of people living there was fixed at four in 1709, and there were four women living there in 1820.

St Catherine's Hospital, by W. Twopenny
This early 19th-century view shows the windmill behind on the Mount
(York Museums Trust (Art Gallery) YORAG:2390)

(34) When Sir John and Leonard Simpson bought it for their business in 1833 they built a house at the back in Holgate Road (at that time known as Holgate or Holdgate Lane) for the almspeople. It was designed by George T. Andrews, who was also the architect of the Victorian railway station in Tanner Row. The four women moved into it in 1835. The old hospital was demolished subsequently, probably in about 1850, and something of its style is said to survive in a curious bay window on the south side of the Kwikfit Garage. It had been customary to appoint widows, but in 1862 single women were made eligible, the stipend being £18 per year.

Bill Sessions visited it in 1946 and found that there was no inscription on the house, and that it was locally known then only as number 45, "even the postman not knowing it as 'St Catherine's.'" At this time it was for women over 60, resident in York for two years. The pensions paid to the four residents came to £72 per year. They each had their own fire, oven, and gas ring; there was gas lighting, but there was no bath or WC in any of the dwellings. There were two sinks with cold taps only, outside in the passage and on the landing. The WC was across the yard. He noted that coal deliveries and bins for emptying had to be brought through the house, and that a separate entrance into the yard would have been very useful to the residents. As so often, he also observed damp. This was caused by houses further up the lane, which were not on main drainage. The adjoining garage produced water too, which flowed down the lane, and also noise and an unpleasant smell of burning rubber. Mrs Waterhouse had lived there for ten years and found doing the washing a problem as there was nowhere under cover. She would have preferred a separate room built off the yard, instead of the "jerry-built air-raid shelter, which they had scarcely ever used (preferring to be indoors)". Another lady, who was short-sighted, was frightened of the gas fittings. The young Bill Sessions also met "a prim & excessively polite spinster upstairs".

As an endowed municipal charity, with five subsidiary charities, it was merged with forty-five others for administrative purposes in 1956. It is now part of York City Charities. The house survived until about 1962 when it, too, was pulled down. The residents may then have moved into Thomas Fothergil Homes. Flats recently built on the site in Holgate Road have been named St Catherine's Court.

(7) Holy Trinity Hospital
The Merchant Adventurers' *Maison dieu* Fossgate

This establishment in Fossgate, was in existence by 1365, when it was already receiving bequests, and through John de Roucliff received its royal licence in 1372 to found a hospital for thirteen poor infirm persons and for two poor clerks to keep their schools. It catered mainly for poor members of the guild and, failing to acquire adequate endowment, it was taken over by the Merchants, or Mercers, Company of York who financed it thereafter. In 1411 a new chapel was built for the master, brethren and sisters, two chaplains, two clerks and thirty sick poor people to celebrate mass. The hospital was situated in the undercroft of the Merchant Adventurers' Hall, from which the sick could see the chapel altar.

Merchant Adventurers' Hall, Fossgate
The undercroft and chapel

The guild and its hall survived the Reformation (and flourishes to this day). It continued to take in the poor, particularly widows in the seventeenth century. In the early nineteenth century it consisted of two rooms housing five men and five women each receiving an annual stipend of £5. In 1879 they were receiving nine shillings a month. At this time the two rooms were described as being dilapidated, and by 1905 had fallen out of use. It had a great many endowments which were turned into a pension charity when the hospital closed in the early twentieth century. In 1931 eight pensioners were being paid ten shillings a month.

(8) St Thomas's Hospital Micklegate Bar
(39) Victoria Bar, Nunnery Lane

Hospitals were often built outside the walls of cities as hostels for indigent travellers or the benighted arriving after the city gates had closed. The Hospital of St Thomas the Martyr (Thomas Becket) stood outside York's Micklegate Bar until 1863. It was founded before 1391 for the maintenance of poor persons of either sex living in the area, and also offered hospitality day or night to all poor travellers, pilgrims and the sick poor passing through York. At one time it was capable of housing twenty-four poor people in dormitories, as well as the brethren of the Hospital. It was transferred to the guild of Corpus Christi in 1478 on condition that it kept seven alms beds, and though the guild was dissolved in 1547 the hospital retained possession of its estates for some thirty years. Ten almspeople lived there at this time, each receiving an annual stipend of 6s 8d. The hospital could also take in eight poor strangers. After that the whole institution was granted to the recorder and town clerk of York which enabled it to continue its work for the relief of the poor. In 1553 it was already under pressure from the many poor who came to its door. In 1574 there were eighteen people in the hospital, including three children. In the seventeenth century it housed about ten people. It was partly rebuilt in 1787, to provide six apartments on each of two floors for aged women who were obliged to beg for alms to maintain themselves. Later, wealthy benefactors like James Luntley and Lady Conyngham (both ubiquitous as charitable benefactors) provided a more stable income. By 1860, however, the hospital was reporting damp on the ground floor, poor ventilation and dark rooms and in 1863 it closed. The building was demolished to widen Nunnery Lane. The hospital appeared in many drawings and paintings of this gate into the walled city and, before it was demolished, also a few photographs.

St Thomas's Hospital, on the corner of Nunnery Lane, in the 1850s
(Evelyn Collection; City of York Libraries. Archives and Local History Dept)

(39) Victoria Bar, Nunnery Lane. This building was erected in 1862 for the residents of the old hospital. In 1906 the eleven almspeople were sharing stipends of abut £80. Bill Sessions seems not to have taken to St Thomas's nineteenth-century establishment when he visited it in 1946. His description contains a good many negatives: "ugly yellow brick", "an iron gate (which sticks)", a small yard "very cramped for drying purposes for twelve people". He noted that there was nowhere for the residents to sit outside away from the noisy traffic at the front, whereas if a wall were taken down there would be space to sit at the foot of the grass bank under the city walls – as well as improving the drying area. The rooms each had an old-style open range and gas lighting; most had a gas ring leading off it, and one had a cooker. Two rooms had a cold-water sink, the rest had to share sinks and draining boards at each end of the passage. The residents also shared a WC, one between six on each floor. There was no caretaker, and one resident was clearly in the early stages of dementia. She had recently lost her husband and had become "queer and dirty" according to the others. "She was certainly queer [meaning odd] ... as she had only been there a few months one wondered at the basis of selection". Another had a talkative 28-year old parrot for company and a few had wireless sets. Several rooms had been empty for some time, one of which needed fumigating and Mr [Charles] Robinson, Clerk of this and several other almshouses, had evidently suggested using it for laundry purposes.

In 1959, it had vacancies for York women suffering from age, or illhealth through accident or infirmity, and who could no longer maintain themselves – but of those, only women who had shown "reasonable providence" would be considered.

St Thomas's 1862 hospital at Victoria Bar

Like St Catherine's it was an endowed municipal charity, and was one of forty-six charities amalgamated, together with its five subsidiary charities, in 1956 as part of York City Charities. The almshouse closed in 1972, when only two residents remained, and was put up for sale. It narrowly avoided demolition at the hands of the city council. Instead it was bought by an entrepreneur and converted into a hotel (the Moat Hotel): an appropriate fate for a former wayfarers' hospital.

(9) St Helen's Fishergate

St Helen's in Fishergate was another leper hospital, which stood near a church dedicated to St Helen, and possibly attached to it. The church and its cemetery were in the area of Fishergate House and Marlborough Grove. A "spittle house" in Fishergate beside St Helen's was in existence in 1399. After the decline of leprosy it accepted anchorites but eventually, in very bad condition in the late sixteenth century, was supporting only a blind man, his family and an old man, all of whom must have been utterly destitute to want to live there. It was demolished in 1622.

Excavation in 1994 of the church's cemetery uncovered the remains of a woman who had evidently been on pilgrimage to Santiago de Compostela because the distinctive scallop shell of the pilgrim was found with the skeleton. The badge of the pilgrim permitted her to beg for alms, hospitality, and exemption from tolls. Perhaps she had been an inmate (or anchorite?) of the hospital at sometime in the fifteenth century.

(10) The Confraternity of St John the Baptist, and The Merchant Taylors' *Maison dieu.* Aldwark

Many of the members of the fraternity were tailors. Their guild hall was built into the bank supporting the city walls in Aldwark in the late fourteenth century. Around 1415 they built a *maison dieu* (referred to in the records as the "mason dew"), for retired or destitute tailors, located somewhere between the hall and the church of St Helen on the Walls. It was one of the buildings round the hall that were pulled down in 1702-3. The long-lost location of St Helen's church was revealed when its foundations were uncovered during excavations on the site of the Ebor Brewery, south of the hall, in 1973, and it seems likely that the present *maison dieu* is close to the site of the original.

The new almshouse was built next to the hall in 1730, directly onto Love Lane, and was a single-storey brick cottage, with four rooms. It may have been designed by the architect William Etty (1675-1734). Each had its own door, with three steps up to it, which opened straight into the room. The ceilings were low, and each room had a range. Its position in Love Lane, as mentioned before, probably lacked charm as it was the route to a latrine on the city walls, itself next to Harlot Hill Tower whose name indicates the nature of the area. Harlot originally meant beggar or vagabond, but even by the fifteenth century had begun to mean prostitute. The surrounding area of St Andrewgate and Bedern became part of a red light district and slum as the buildings of the former Vicars Choral decayed.

The Merchant Taylors' former *maison dieu*, Aldwark

Hargrove (1818) suggests that a room in the hall was also used to accommodate another poor person. He states that the stipend ranged from £2 10s to £4 per annum. Some out-pensioners were also supported by the guild, and three other tenements were rented out in the nineteenth century. In 1909 the residents received about £14 a year, supplemented out of Company funds to make it up to £40 a year. The accommodation was still for former tailors or their widows who were not receiving poor relief, being still able to make a little income from their trade.

By 1946, when Bill Sessions visited the almshouse, it was already being used by the caretaker of the Merchant Taylor's Hall and her brother. In 1962 it was converted into a pleasant cottage for the caretaker to live in. Three of the doors were bricked up and made into windows. Part of the cottage was converted into a kitchen, and a connecting passageway was constructed across the old Love Lane, closing off the route to the city walls (and the latrine on the walkway). There is still an inscribed panel on the front of the house, which says "This hospital, built by the Merchant Taylors Company, for the use of four poor brothers or sisters, in the year 1730. Mr Thomas Spooner, Master, Mr John Napier, Mr John Etherington, Thomas Ellis Sen., & John Dunning, Wardens."

(11) St Anthony's Hospital, Gillygate/Horsefair

This hospital started as a chapel of the great hospital of St Anthony of Vienne (Isère, France), and was located near the Horsefair, probably at the junction of the present Wigginton and Haxby Roads. This would have been agreeably close to the Dean and Chapter's gallows. It was vacant in 1401 when a hermit moved in and collected alms ostensibly (and illegally) for the repair of the highway. He was evicted in 1403 and a hospital was established by about 1420 when it held a matron, a married couple and two other people. Indulgences were granted in 1429 to anyone who gave alms for its support (Calendar of Papal Letters, vol VIII, p. 85).

There is also mention of a hospital dedicated to St Anne in the area of the Horsefair in RCHM v.4, but no evidence for its existence is supplied. It may have been a chantry chapel which took in the indigent.

(12) Cordwainers' *Maison dieu.* St Denys Church Lane

The house was originally founded by John Marton, a member of the Company of Cordwainers, before 1436 in a lane near St Denys's church. The Company was dissolved in 1808 and the *maison dieu* and the archives of the fraternity were transferred to William Hornby, the last master of the Cordwainers' Company. He was a boot- and shoemaker in Blake Street, and also Sheriff of York in 1807-8. He rebuilt the house in 1811, probably on the same site, as four comfortable dwellings for destitute shoemakers. Occupants had a room on the ground floor and a bedroom above for which they paid one penny a year. Baines (1823) mentions that there was formerly a cupola and bell which tolled for the death of any of its members. On the death of William Hornby, the houses passed to his son, the Rev. Surtees Hornby.

The owner in 1861 was found to be charging a full rent, and as a result the *maison dieu* was taken over by the York Charity Trustees to provide once again for shoemakers and their dependents. In 1909 it was still able to house four poor cordwainers of good character who, from misfortune, were unable to maintain themselves. The cottages were reportedly poor and damp in 1912, and were finally pulled down in 1938.

Baines (1823) says that the cottages were located in St Dennis [sic] Church Lane; they appear on the 1852 map in Church Lane, which later ran into the bottom end of Piccadilly. Elizabeth Brunskill (1960) gives the location of these "two demure cottages with their little gardens" in the 1930s as "near the Selby bus-stop in Piccadilly". Bill Sessions, in his manuscript notes on York charities in 1949, mentions a conversation with Mr Swaby, secretary of the Tadcaster Tower Brewery Company, which was on the left past the Employment Exchange in Piccadilly. The company had bought "some tumble-down cottages fronting on Piccadilly at the left-hand side as [you] go into the main entrance of the company. He believed one cottage had a cobbler's bench in it." Bought for site value, they were pulled down, "and are now a small grass square ... surrounded by a brick wall adjoining the road." The area has been developed with hotels in recent years and the site has virtually disappeared.

Unless it is merely coincidental, the location and the date of demolition make it possible that it was this property, or part of it, which was acquired in an exchange between the parishes of St Crux and St Denys in 1895 and that these cottages were therefore partly the St Crux Parish Almshouse (**21**), which had moved from Fossgate. Marion Marton (wife of John?) who died in 1444 was a parishioner of St Crux church.

A pension charity was created out of the endowment and, as the Cordwainers' Pension Charity, continued to be managed by the York Charity Trustees, until in 1956 it became part of the group of forty-six such charities administered under one Charity Commission scheme as York City Charities.

(13) St Anthony's Hall Peasholme Green

St Anthony's Hall in Peasholme Green was built after 1446, when King Henry VI granted a charter to found the guild of St Martin on the site of a former chapel of St Anthony. The new hall had a hospital chapel for the use of seven poor men and members of the guild. The almspeople lived on the ground floor near, or in, the chapel. After the Dissolution the charity continued under the government of the Corporation of York, and by 1569 it was in use as a poorhouse where the inmates undertook weaving to make an income. After that it was variously used as a house of correction, a knitting school, a Civil War ammunition store, a military hospital, a prison for vagrants unable to find work and illegally living away from their own parish, and a charity school – the Blue Coat School. As part of the University of York from the 1960s, it became the Borthwick Institute of Historical Research, housing local and diocesan archives. In

2005, the Institute and the archives moved up to the university library in Heslington and the hall was bought by the York Conservation Trust in 2006. Excavated and restored, it now houses the Quilt Museum and Gallery.

St Anthony's, Peasholme Green
From the garden

(14) The Common Hall Coney Street/Lendal

The Guildhall off Coney Street, once a hall in common use by several guilds, has long been used by the civic administrators of the city of York. It was mentioned first in a charter of 1256, but the present hall was built in the mid-fifteenth century. Badly damaged by bombing in 1942, it was restored after the war and is still in use by the city council. To the east of it originally stood its chapel and *maison dieu*, dedicated to St Christopher. These were previously sited in Fishergate Within (now Piccadilly) probably very close to the Cordwainers' *maison dieu* (**12**), and moved here in 1456. The guild of St Christopher was dissolved in 1549. By the 1720s the chapel had become a grubby public house called the Cross Keys. It was demolished in 1725 and the site of both chapel and *maison dieu* (which may have been under it) now lies beneath the Mansion House.

(15) St Mary Magdalene Clifton

This establishment was founded possibly as a leper hospital given its dedication to Mary Magdalene and its distance from the city walls. It was a chapel and hospital situated in Clifton, near the city boundary on the east side of the junction with the present Burton Stone Lane, an earlier name for which was Chapel Lane. The refoundation by John Gysburne,

Precentor of York who died in 1481, was for two chaplains and it also supported retired clerics.

According to Rotha Clay, who appears to call it "St Mary the Less" (a dedication to the Virgin), this hospital was founded in 1481 probably near the Horsefair, but like, or because of, William Hargrove she may have confused two different establishments, this one and **(5)** above.

(16) Sir Thomas Hesketh's Hospital, Heslington

Sir Thomas bought the mills next to York Castle in 1603, intending to use the income from them to found and maintain a hospital for retired workers on his estates in Heslington. He died before this could be carried out, but a hospital was established by his wife Dame Julia in 1608 beside the main road north of Heslington Hall. Eight almsmen lived there, and possibly one woman. In 1630 the hospital received £50 a year from the Castle mills endowment. Its master was paid £6 13*s*. 4*d*. a year and the inmates received £5 each. There was further endowment of £5 a year from land at Hutton Rudby in the North Riding.

In 1642, during the Civil War, the Castle mills suffered a bad flood which wrecked the dam. This was followed by one of the mills being set to grinding gunpowder – with predictable results. The mills were destroyed in the blast and as a result no rent was paid for the benefit of the hospital. The master, Richard Kendraw, petitioned for years for the mills to be rebuilt and for financial aid. He received ten pounds from Colonel John Bellasis (Governor of York), but the mills were not restored until 1649. The income from the mills yielded £45 in 1892, at which time it was also receiving two other annual endowments, of £5 and £20, from Lady Amherst's estate, and from the Squire, Robert Bateson de Yarburgh, who as Robert de Yarburgh-Bateson later became 3rd Baron Deramore, founder of the other almshouse in Heslington **(43)**. Though the endowment continued to provide a small annual sum for a time (paid by York Corporation as the rent-charge on the mills), the residents now pay a maintenance contribution.

Sir Thomas Hesketh's Hospital, Heslington

The hospital was moved to its present position on the road to Fulford in 1795 by Henry Yarburgh, the squire. The new rector of Foston, Sydney Smith, encountered him in Heslington in 1809, and wrote a famously entertaining description of him (*see* Pearson). The rebuilt hospital is a two-storey building in brick and flat tile, with a central pediment. There was originally a chapel in the centre with, either side, four two-roomed apartments at the front for men, and a room behind for a woman. The foundation tablet set into the bricks under the pediment contains a quotation from Psalm 68:10 "Thou O God hast of thy goodness prepared for the poor", which was also used in the foundation tablet of Lady Hewley's (**26**).

In 1968 it was repaired and modernised by Lord Deramore and remains a functioning almshouse. Several doors and windows were blocked when the cottages were joined together to make four two-bedroom maisonettes, and two one-bedroom flats for people of limited means. Its roof is under repair in 2010.

Since 2010 it has been part of the Heslington Almshouses Charity, together with the Lady Deramore Memorial Almshouses (**43**) in Heslington Village.

(17) Watter's Hospital Chapel Row, Walmgate

Sir Robert Watter was twice Lord Mayor of York. In his will, proved 15 June 1612, he left money for the establishment of a hospital in several houses that he owned in a lane off Walmgate, for the perpetual maintenance of ten persons. The name of the lane is spelt in innumerable ways, including Neutgate, Newtegate, and Nowtegail Lane. It seems to have been a narrow lane leading from Walmgate to Fishergate Bar. The name, which appears to mean "no-way lane", suggests it was a cul-de-sac. It was later widened to become George Street. Sir Robert seems to have wanted to erect a new hospital but, though the existing cottages remained in use and were repaired in 1627, no new accommodation was ever built on this site.

Watter's Hospital, Nowtegail Lane
Drawing by George Nicholson, 1825 (*Evelyn Collection*)

A governor or reader was to be appointed, together with certain brothers or sisters who would be paid to manage the hospital. Such readers and ministers sometimes looked after the conduct of several institutions. These appointments often failed for lack of money, and hospitals were frequently left to fend for themselves.

In 1823 it had become customary to appoint only poor women. William Hargrove's efforts to get the arrears owing to the hospital paid, and the women helped, were acknowledged by the Brougham Commission which was gathering information about financial mismanagement of such institutions at that time.

Even by 1818, the timber-frame buildings were deteriorating, and in 1844 when George Street was constructed they were pulled down and a new hospital was built at the end of Chapel Row, which leads off it. This building appears on the 1852 map. It was a tall two-storey building of brick and tile construction, making up nos 5-11 Chapel Row. Three had doors onto the street, with access to the rest from the back via a covered passage. It housed seven people in one-up, one-down accommodation, placed back to back. Each of the residents received £2 2s a year, with a ton of coals in addition. These had to be respectable old people with some means of their own, such as a pension from a former employer. The hospital also supported three out-pensioners. From 1925 it was administered by York Charity Trustees.

Watter's Hospital, Chapel Row
(City of York Libraries. Archives and Local History Dept)

Bill Sessions found it to be very damp in 1946, with brick floors covered in linoleum, the underside of which was regularly covered with "black wet". The living rooms of each had an open fire and oven, with gas lighting, a gas cooker, and also a gas poker. A stair led up to the bedroom with a "little-used fireplace". None of the rooms had running water. Instead there was a tap outside at the back of the dark passage leading to the stairs, which were lit by a gas jet. The yard was in bad condition, partly concreted but often wet. There was a scrap of garden carefully tended by Miss Gledhill, the youngest of the seven women living there. Across the yard were two WCs shared by all the residents. A coal delivery was made during Bill Sessions' visit, and though the coalman was careful inevitably there was a mess of coal dust which upset the sick old lady living there. Another lady, Mrs Fletcher, had lived previously in a modern house and took the prevailing conditions badly; she feared that another winter in the damp would see her off.

Its income in 1955 was £32. As it was already in very poor condition, and as it was a site which was thought might be useful for the Roman Catholic school nearby, it was compulsorily purchased and demolished in about 1958 (it is now a carpark). The sum raised and the charity's endowment were managed by York Charity Trustees until the 1956 Act which merged forty-six endowed charities under one Charity Commission scheme. It is now administered as part of York City Charities.

(18) Agar's Hospital Monkgate

Thomas Agar had three cottages (six rooms in total) built in Monkgate early in the seventeenth century. In his will of 1631 he directed that the ground behind them should be laid out to provide a garth – a garden or vegetable plot – for each house, and that six poor widows, from the parishes of St Maurice or St Michael le Belfrey, should be appointed to occupy the cottages. The cottages were burned down during the Civil War siege of 1644, and were rebuilt. The residents, only a few minutes walk from Monk Bar, would no doubt have taken shelter inside the walls of the city.

Agar's Hospital. Drawing by George Nicholson, 1825
(*Evelyn Collection*)

In 1818 the widows were each receiving a half-yearly stipend of £1 13s 4d, plus 5 shillings from the rent of the adjoining field (Hargrove 1818).

The cottages were still in good condition up to the early nineteenth century, when family members were still among the trustees, though there were concerns that the roof would probably need replacing fairly soon.

These members of the Agar family were Catholics and they appointed only Catholics to the hospital. The cottages appear on the 1852 map, and in *The practical guide to the City of York* of 1880, by which time they had been certified as unfit for occupation and sold. By 1895 Agar Street had been constructed over the site. The hospital was replaced by a pension charity for six poor widows.

(19) Ingram's Hospital 90 Bootham

There is a good deal more known about this hospital in Bootham than most. It is one of the most attractive buildings in a remarkably beautiful street. It survived virtual destruction during the Civil War siege of 1644, and also the poor condition it fell into in the twentieth century.

Sir Arthur Ingram was a wealthy and influential man, Secretary of the Council of the North and owner of, among other large mansions, Temple Newsam House. He also had a house built on the former city palace of the archbishops behind the Minster, where he entertained Charles I in 1642 when the King held court in York. He was four times MP for York, and High Sheriff of the county in 1620. In 1629/30 he acquired land for a hospital in Clifton from Thomas Sandwith, and his hospital was finished in 1632 with provision for ten poor widows. Built of dark red bricks with stone facings and a tiled roof, it is dominated by a low central tower over a former chapel and caretaker's rooms. The decorative central doorway, of c1190, was bought from Holy Trinity Priory, Micklegate for 13s 4d (a mark, or 2/3 of a pound) – its provenance is recorded in the Ingram accounts at Temple Newsam, though there was once an alternative theory that it came from St Giles' church, formerly in Gillygate.

The hospital was badly damaged, probably gutted by fire, during the 1644 siege and needed rebuilding. In 1649, five thousand bricks were needed for the structure, and all timberwork, including the roof, floors and staircases, was renewed. In the 1970s when work was being done on the roof of Ingram House, the contractor (the father of John Redhead of JLR Roofing) found a seventeenth-century bill for post-Civil War repairs stuck into the rafters. His story, hitherto regarded as myth in the Redhead family, is likely to be true.

The accommodation consisted of living rooms measuring 12x18 feet, with an open fire; a narrow staircase led to the unheated bedroom. Outside were privies and a well. The garden contained allotments for the residents. There was a chapel on the back of the house, which was converted before 1900 into a dwelling, and its end window filled in.

There was a regular stipend, which continued to be paid after 1644 so it can be assumed the ladies were safely ensconced elsewhere for the duration. In addition they were given materials for a new gown every three years, and a silver-gilt badge bearing the Ingram crest, a cock. The occupants of the hospital were able to augment the stipend by hemp spinning.

Bill Sessions visited the house in 1946 and described conditions there, which were seemingly unchanged from a previous age, and quite unacceptable in the twentieth century. The house was lit by gas, and there were large and ancient open ranges with huge chimneys "the width of the wall – so that boys could [in the past] get up to sweep the soot". The chimneys smoked badly and made the rooms dark. Outside at the back were four WCs and each resident had an outhouse for laundry with sink and cold tap "which have to serve all needs". There was no bath. The rather bleak (unheated) bedrooms were reached by "an extremely dangerous narrow stairway built in the chimney wall." The caretaker's dwelling in the centre was, like the rest, picturesque from outside but awkward within. Slates or tiles lost during the 1942 air-raid had not been properly repaired so the roof leaked. Her sitting room occupied the former chapel. She was responsible for looking after the enormous garden, part of which was let to a local butcher to keep chickens.

The caretaker's sisters-in-law (the Misses Gunney "of uncertain age & talkative") had been brought up in the house, when Mrs Meynell Ingram of Temple Newsam still had control. She sent families of deceased farm workers to live there to free up the tied cottages on the estate. Conditions then had been even more primitive, with two pairs of earth closets in the garden. These ladies disapproved of the laxity of the present trustees who were letting in a lot of Dissenters "even a Quaker, Mrs E. Florence Baker". Victorian prejudices were very long lived.

Ingram's Hospital, Bootham
Central tower

In 1957 it was taken over by Ings Trust (John Bowes Morrell and his brother Cuthbert) and converted into five flats, providing more windows in the south-facing back walls but incidentally removing much of the original interior. It was one of the many properties bequeathed by the Morrells to the York Conservation Trust. The flats, two on each side of the central tower, and one occupying the tower and former chapel, are rented out to private tenants. John Bowes Morrell provided land behind his house, Burton Croft in Burton Stone Lane, for a new almshouse range to be built for the accommodation of the almspeople. Burton Croft itself was pulled down in 2005 despite strong opposition, to be replaced with close-packed Barratt Homes.

Ingram's Hospital Charity became one of the forty-six merged in 1956 under single administration as York City Charities.

(20) Anne Middleton's Hospital Skeldergate

Mistress Anne Middleton founded a hospital in Skeldergate in 1659, on ground leased from the Vicars Choral. She left £2000 in her will to the Lord Mayor and Commonalty of York for the building of a house for twenty poor widows of freemen of the city. Her husband Peter was Sheriff at the same time as Thomas Agar was Lord Mayor. In 1771 the front was taken down and rebuilt about a yard further back – this house is depicted in a drawing in the Art Gallery. A further scheme to rebuild came in 1815, and was undertaken by the Corporation in 1827-1829. The new building, designed by Peter Atkinson Jnr, was erected considerably further back. It consists of seven bays, the central bay projecting under a pediment, with a niche above the present front door in which there is a stone figure representing the foundress in seventeenth century costume. Inside there were eleven rooms on each floor: six at the front, five at the back, with a central staircase. When the Joseph Terry Homes **(42)** were built in front of the Hospital, entrance to Middleton's was through a gate into Carr's Lane.

Anne Middleton's Hospital, Skeldergate
Now a hotel

Bill Sessions was fairly impressed by conditions here in 1946. The hospital became part of the Municipal Charities Trust in 1939 and the interior was remodelled and enlarged to provide accommodation for ten almspeople and a warden. Before this, there had been eighteen almspeople each with only one room. Now the eleven had two rooms each, a bedroom and living room, with electric light, a fire and an oven. There was still only a sink with cold tap, however, and for two of the dwellings that was in the passage outside. There was now an inside WC on each floor leaving the two outside WCs little used. There was also a bath on the first floor, the water heated by a slot-meter gas geyser. There was damp at one end of the ground floor, the resident in consequence having to sleep in her living room. However, the house evidently had a cheery atmosphere. At that time the residents received about two shillings a week stipend.

Though it was kept in good repair, by the 1960s conditions were not ideal. It was closed in 1973 and sold for about £34,000 to Andrew Clark and his wife Käthe who turned it into a hotel. The Hospital's endowed charity, which included nine subsidiary charities, was one of the forty-six amalgamated in 1956 to be administered as York City Charities.

(21) St Crux Parish Almshouse Fossgate (later moved to Piccadilly)

This almshouse was founded by the will of Edith Darke in 1663. In 1825 it consisted of two rooms off a yard in Fossgate which were occupied by two poor widows of the parish. A new almshouse seems to have been acquired in Piccadilly in about 1895 by an exchange with St Denys' parish. It was compulsorily acquired by York Corporation in 1935 and the building demolished. The similarity of location and date of demolition suggest that it may share its identity with the former Cordwainers' *maison dieu* **(12)**.

(22) Ann Wright's Almshouse Bishophill

This almshouse was established in the parish of St Mary Bishophill Junior, by Ann Wright's will of 1670, which provided for a freeman's wife or daughter. The building was replaced in 1768 and by 1825 was no longer in use as an almshouse, though the trustees were recommended by the Brougham Commission to let it to poor freemen's widows or daughters. It does not appear on the 1852 map as an almshouse, therefore, so its exact location is not clear. The charity was consolidated with other charities of the parish in 1907 (Consolidated Charities).

(23) Poor's Cottages Front Street, Acomb

The identity of this poorhouse remains obscure. There was a row of cottages on the south side of Front Street, Acomb, built probably for Manor Farm workers at the end of the seventeenth century. At some time in the nineteenth century they were pulled down and replaced with cottages grouped round three sides of a rectangle. They have apparently had several names, including Whitehall Cottages. They seem to have been taken over for use as poorhouses, originally vested in the Guardians of the Great Ouseburn Union, and let by the overseers (and therefore were not almshouses). The rents were applied (i.e. put to use) in aid of the poor rate. If these were the cottages involved, they were known to the Charity Commission as Poor's Cottages, and to Acomb Parish Council and the Great Ouseburn Rural District Council, as Town's Cottages, and Ogram's Yard. They were bought in 1948 by York City Council. Harold Richardson (1963) records that they were condemned as insanitary and pulled down in about 1950. The site was where the library was first built in the 1960s.

Whitehall Cottages, Front Street, Acomb
(Acomb Local History Group)

White Row, Front Street, Acomb

According to Harold Richardson's map, however, Ogram's Yard was behind Gale Farm at the end of Front Street. There is still a row of cottages nearby, also on the south side of Front Street, known locally as White Row, which have often been confused with the former poorhouses. They too were supposedly built as copyholders' cottages (a form of tenure provided on big estates in return for agricultural service) in about 1691. They may have been for agricultural workers at Gale Farm but it is not clear how they were used later nor what link they had with Poor's Cottages. They may also have been owned by one Francis Barlow in the nineteenth century. They were modernised in the 1920s, and the internal measurements still correspond to details in seventeenth-century Acomb Manorial records. This row is now attached to Chancery Court, a new residential estate for the elderly.

(24) Sir Henry Thompson's Hospital Castlegate

Sir Henry Thompson founded this hospital in 1692. It was built in 1700 in Castlegate, next to St Mary's Church, for the benefit of six poor men of the parish. On the front of the house was an inscription quoted as: "This hospital was founded by Sir Henry Thompson of Middlethorp, sometime Lord Mayor of this city & Dame Anne his wife for the relief of 6 poor men : Anno Christi 1700". The house was repaired and refronted in 1813. Financial support came from investments in Consols and, in the early twentieth century, sale of land next to the barracks to the Ministry of Defence. The almspeople each had two rooms and, at the end of the nineteenth century, a fixed annual stipend of £20. There were in addition six out-pensioners receiving £13 in 1934. By 1937 the hospital was ordered to be demolished apparently to make way for new municipal offices.

Demolition was suspended during the war and not carried out until 1949. When Bill Sessions visited it in 1946 there was still one solitary inmate living there in slum conditions. He was a "shuffling old gent named Mr Hallett with an old cap and a pair of blue eyes which didn't follow each other". His room was off a dark stone-flagged passage, and had a small range in the corner. Lighting was by gas and Mr Hallett, having the place to himself, had tipped a supply of coal in the side passage. Water came from a primitive tap in the yard, where there were also two former WCs which had ceased to function and were foul. A staircase led to the unheated bedroom through a boarded off area in the living room, arriving in the middle of the bedroom floor. The old man's bed-linen was "distinctly off-white". Hearing the sound of someone in the passage, and finding a youngish man there "ostensibly unlocking his bicycle", Bill Sessions gathered that Mr Hallett had been subletting one of the rooms to make a bit on the side – which accounted for his being able to afford the two bottles spotted by the bed.

The hospital appears occasionally in photographs of St Mary's church, Castlegate. After its closure the almshouse was reorganised as a pension charity. As a corporation charity it had been administered by York Charity Trustees since 1837, and in 1956 was one of the forty-six endowed municipal charities merged to form York City Charities.

(25) Barstow's Hospital Blossom Street
(38) Caroline Street/Clement Street.

According to William Hargrove in 1818, the tradition was that this hospital was founded by two maiden ladies called Bastow or Barstow. It seems to have been founded by Alice Barstow (actually a widow) who died 25 April 1724, and perhaps her sister Elizabeth who died 21 November 1747. Their wills do not mention the hospital, which may have already been transferred to other ownership. They gave some old cottages in Blossom Street in May 1702, stipulating that six poor women should be given separate and convenient rooms.Its original position in Blossom Street is marked on the 1852 map of York, just before the right turn into Holgate Road. In 1856 it was occupied by five poor widows. In 1859 the house had become inferior and insanitary and permission was requested to sell. It sold for £450, and a new site was purchased for £47 in Caroline Street.

Barstow's Hospital, Blossom Street
Drawing by George Nicholson, 1825 (*Evelyn Collection)*

(38) The new almshouse cost £350 to build, on its new site on the corner of Clement and Caroline Streets, at the Clementhorpe end of Nunnery Lane. Its foundation stone was laid by Mr B. Barstow. The house was of brick with stone window facings and a slate roof, containing room for four women. Bill Sessions' photograph (p.144) shows a dramatic pediment over the front door. A dark passage led from the front door to the four bed-sitting rooms.

Those selected to become beneficiaries had to be poor, infirm women of good character, resident in York, over sixty (unless disabled), and not receiving poor-law relief. There were four frail old ladies living there in 1946 when Bill Sessions visited, who were all terrified of becoming helplessly infirm because then they would have to move out, and the prospect of the workhouse hung over them. Shades of Betty Higden in

Dickens' *Our mutual friend*, even at this late date, though one commented that it was said to be very nice once you got there. The ladies received quarterly payments, in one case of £2 1s 0d.

Lighting was by gas, and each room had a small range with an oven, though some had installed a small gas cooker in the living room, and one had a gas poker for firelighting. There was one shared cold tap in the passage, and one shared indoor WC. The rooms had no sink, only a little annexe serving as a food store with a shelf for a washing bowl, and in two cases a coal store. Coal was mostly kept in one of the outhouses in the yard.

Elizabeth Brunskill recorded the almshouse as still in occupation in 1960, but it was demolished, probably after the last resident had left or died, in 1967, along with the whole area of terraced streets. There is only a very short stretch of Clement Street left. The charity was merged with Cameron Walker Homes, but there is no record of any remaining residents moving there. The site is now under a housing estate, but the name Barstow House has reappeared, very suitably, on some local authority sheltered housing nearby.

(26) **Lady Hewley's Hospital**, Tanner Row
(35) St Saviourgate

Lady Hewley's Hospital was originally founded in Tanner Row in 1700. Sir John Hewley was MP for York in the late seventeenth century, and a parliamentary colleague of Sir Henry Thompson. He and his wife Sarah, Lady Hewley (with whom he had eloped), were non-conformists and frequently sheltered George Fox, founder of the Quaker movement, at their house, Bell Hall, in Naburn. They also had a house in St Saviourgate where Lady Hewley preferred to live after the death of her husband in 1697. She was a wealthy woman with no living children; her charitable endowments were and remain very valuable. She contributed to the cost of construction of the chapel nearby – though after her death in 1710 she was buried in St Saviour's church. She set up two charities, one to offer financial support to preachers or their widows, and the education of young men for the ministry; the other to support her almshouse in York and the people living in it.

The original almshouse, or hospital, was built at the Black Fryars, the site of a Dominican Friary in Tanner Row, opposite the end of Barker Lane. Thomas Jeffrey's 1772 plan of York shows its layout (RCHM v. 3, p.xxxv).

It sheltered ten poor persons, nine of whom would be poor widows or spinsters over the age of fifty-five, and preferably dissenters. The tenth would be a man who would lead their prayers twice every day – or a woman if a man could not be found. To ensure the almshouse-beneficiaries were of good character, they were required to be able to recite the Lord's Prayer, the Creed, the Ten Commandments and the, now somewhat obscure, catechism of Edward Bowles (who had helped to preserve the Minster and its library after the 1644 siege). The widows received a stipend and fuel. They were not allowed to beg, though they were allowed to accept freely-offered alms and to sell articles of their own manufacture on the street.

The house was of brick construction, with wings at each end enclosing a small courtyard at front and back, with a chapel at the rear. The donor's arms were set over the entrance. The house was well maintained by the trust, and the stipends (increased from time to time) continued to be paid to the residents. In 1826 it was decided to pipe mains water from the Water Works Company into the hospital for the benefit of the poor people living there.

Lady Hewley's Hospital, Tanner Row
Engraving (*Evelyn Collection*)

In the 1830s George Hudson and several York councillors were involved in bringing the railway to York. George Hudson in particular wished to build a station inside the city walls. The required area, including Lady Hewley's Hospital, was compulsorily purchased by the railway company, the sum paid for the hospital being £5,105. Though this sounds draconian, it seems that the new almshouses in St Saviourgate were ready for the residents to move into before the old almshouse was demolished, and the railway used a temporary station outside the walls until the new station could be built in Tanner Row. It is possible that the Charity's willingness to accept a generous cash offer to enable them to build bright new almshouse cottages also enabled them to make a successful transition in 1840 without having to find interim accommodation. After a long dispute ending in 1848, people belonging to Baptist, Congregationalist, and Presbyterian Church in England denominations were permitted to be trustees and residents.

(35) St Saviourgate. The cottages, designed by J.P. Pritchett, were built on or near the site of Lady Hewley's townhouse, next to St Saviour's church. They are built of brick with sandstone facades, and slate roofs. The supervisor/caretaker's house stands at the entrance; it has two bedrooms, living room, entrance hall and kitchen. Set into the end wall overlooking the street is the tablet recording the original foundation, with Lady Hewley's arms, brought from the old hospital. It also contains the same quotation that appears in Sir Thomas Hesketh's, from Psalm 68:10, "Thou O God hast of thy goodness prepared for the poor".

There were originally seven cottages and a chapel, but the chapel was converted into another dwelling in 1975 and there are now eight one-bedroom cottages, and two two-bedroom cottages for protestant men and women over sixty. The eight women living there in 1909 received £2 a month each; the lone man received £2 16s 8d – unquestioned in 1909, it is not now obvious why.

Lady Hewley's Cottages, St Saviourgate

Each cottage has a back yard, with formerly an outside WC, and access to a lane. The top end of the lane is now blocked off and is owned by the house next door. The bottom end has no exit so dustbins are brought through a gate into the garden and out by the narrow gate into the street. Each pair of cottages has a common entrance area which by 1946 had been divided to provide separate entrances. The wooden partition is removable for taking furniture in and out. There used to be steps down to a cellar from the entrance, subsequently boarded over and a coal store provided under the stairs. On the ground floor there is a sitting room and kitchen, with formerly a pantry, gas cooker, and sink with cold tap. Above is the bedroom (still with fireplace), a boxroom and a cupboard. In 1946 lighting was already by electricity. The only shortcoming Bill Sessions observed was the lack of a bath – also deplored by two ladies he spoke to. The two cottages at the bottom overlook the Hungate area, formerly a bad slum which was pulled down in the 1930s. In 1946 it still looked like "a blitzed town" and was awaiting reconstruction – as part of it still is in

2010 – as well as the construction of the Stonebow.

New bathrooms and kitchens were built onto the back of the cottages in the late 1950s, but as the bathroom had to fit in with the sloping line of the roof, it involved two steps down from the bedroom, which was hazardous for old people. In 2010, with designs by Richard Parkin, the kitchens and bathrooms are being completely rebuilt and refurbished, two cottages at a time. The old ranges in the kitchen remain, but concealed behind boarding. The outside WC is incorporated as a utility room. The bathroom floor level has been raised by pushing out a small gabled roof extension in order to raise the ceiling level. In addition secondary glazing has been installed. The results are not only welcome but very attractive.

(27) **Dorothy Wilson's Hospital** Foss Bridge

Mistress Dorothy Wilson founded a hospital in 1717, in her own house on Foss Bridge. It was for ten poor women, with preference being given to retired domestic servants. Two schools, one for boys behind the house and one across the river for girls, were also endowed by Mistress Wilson. She gave lands at Skipwith and Nun Monkton as an endowment. The almshouse was rebuilt in 1765, when it seems to have included a schoolroom for the twenty poor boys, and again in 1812 following the rebuilding of the bridge. It is of brick and consists of five bays with, on the street-front ground floor, three arches over the door and windows on either side. A tablet at second-floor level above the door reads "This hospital and school house were endowed by Mrs Dorothy Wilson Spinster for the Maintenance of ten poor Women as also for the instruction in English, Reading, Writing and Clothing of twenty poor Boys for ever. She departed this life 3 November 1717". The endowments produced a generous income and the schoolmaster was paid £20 a year, increased later to £30; the boys were provided with clothes annually and £6 to place them in apprenticeships. Three blind pensioners received £2 each, and a schoolmistress was paid £2 a year to teach six children in St Denys's parish to read.

The house, schoolroom, and schoolmaster's house are marked on the 1852 map of York. In 1909 there were fifteen residents, poor women over sixty, either widows of tradesmen or spinsters who had been in domestic service in York or nearby. Ten received stipends of £16 a year, the other five had to find income of their own.

Even in 1946 Bill Sessions thought it seemed to be in poor structural condition. There was a caretaker's house at the back, looking over the river Foss. This had two downstairs rooms, and three small bedrooms, but no bath. The main house consisted of thirteen bed-sitting rooms on three floors. Each contained an open fire range with oven and there was electric lighting throughout. There was a shared WC, cold tap and sink on each landing, together with individual lockable store cupboards. Coal stores for the house were in the basement, down a very dark staircase – a very long climb with a bucket for old women living on the top floor. The school on the ground floor had been let as a warehouse to Mr Harwood next door. In 1947 the residents were receiving a stipend of £2 12s 6d per quarter.

Dorothy Wilson's former hospital, Foss Bridge

The house was converted into self-contained flats in 1958. It became unstable in the 1990s, and was empty by 1993. It was sold in 1999, charity income being insufficient to pay for the necessary repairs. The charity is now likely to merge with Ellen Wilson's, a nineteenth-century foundation in Lawrence Street. It hopes to build a new almshouse to fulfill the requirement of the permanent endowment.

(28) Colton's Hospital Tanner Row
(45) Shipton Street

Colton's Hospital was founded in Tanner Row in 1717, the year Dr Colton's wife Mary died. Dr Thomas Colton was a presbyterian dissenter; he was also Lady Hewley's chaplain and one of her first trustees. He graduated as an MD in Leyden in the Netherlands, and is thought to have resided at Hewley's almshouse for a time. His hospital was built at the junction of Tanner Row and Rougier Street, and consisted of a row of low brick buildings with a back yard and an orchard. Lands at Cawood and Thorpe Willoughby formed the endowment. It offered a single room each for eight poor women, with no other limitations on selection – the link to the Unitarian church did not produce a problem as it did at Lady Hewley's in the nineteenth century.

In 1909 each resident received a stipend of 12s 6d a month. The report to the Royal Commissioners on Poor Laws and Relief of Distress mentioned that the rooms were poor and damp. It also reported that one of the trustees visited the residents once a month, so conditions were not ignored.

Colton's Hospital, Tanner Row, in 1909/10
(City of York Libraries. Archives and Local History Dept)

(45) Shipton Street. Street widening condemned Colton's Hospital, as it did St Thomas's at Micklegate Bar. The almshouse was demolished in 1910, and the almspeople moved to Shipton Street, off Burton Stone Lane, where eight cottages were built for them. Four more were added in 1930. They are placed round a grassed square with flower beds, two of the newer houses facing the street, the other two at the back, completing the square. Each cottage has a living room, bedroom, kitchen and bathroom, though the kitchens and sitting rooms of the later cottages are divided only by a counter.

Colton's Hospital, Shipton Street
1930s cottage on left, 1910 cottage on right

In 1946 there were no baths, even in the newer cottages, though there was an indoor WC, and a pantry with a sink and cold water tap. A bomb had fallen on a nearby terrace in 1942 causing damage to windows and plaster. Strips of garden at the back were cultivated by a local jobbing gardener if the resident could not do so. Bill Sessions was concerned for the welfare of the old ladies he met one lunchtime in April 1946. There was no sign of any mid-day meal preparation and it seemed that some needed more care and help than was available to them. He found one or two pathetic – "a frail grey lady who could only hobble with difficulty – she had had nine children 'and only lost one'"; another was a widow; left with three small children, she had earned enough by charring to keep them, "sometimes leaving the children with a neighbour & sometimes leaving them 'to providence'".

Preference seemed to be given to Unitarians, even though there was no official discrimination. One lady, Mrs Golden, thought she had had to wait longer for a vacancy because of being "church" rather than Unitarian.

The almshouse borrowed £127,000 on a twenty-five-year loan from the Housing Corporation in 1982, to replace bathrooms and kitchens. It thereby became a registered social landlord and a housing association. The architect of the scheme was David Strickland of the Design Unit, King's Manor, the contractor Hare & Son of Kelfield. When the loan is finally paid off, the charity plans to deregister. Central heating was installed in 1996 using a grant from the National Lottery Fund. Each cottage now has a central heating boiler in its former pantry. Baths are being replaced by showers, which are easier for the elderly to use.

The present trustees have not only to be members of the chapel in St Saviourgate, but also to become trustees of it. Plans to open the trusteeship to a wider group have been successful, and it may soon include members of St Luke's church (the nearest to the almshouse), and also people offering a wider expertise. There is no caretaker, and the trustees make themselves available (perforce) to requests for help from the residents of the almshouse. One, Alfred Fletcher, does all the general administration. Hitherto the clerk, a solicitor, has only prepared the annual Charity Commission return.

(29) Winterscale's Hospital Walmgate

Robert Winterscale's (or Winterskelfe's) Hospital was founded in Walmgate in 1705 by Percival Winterscale, and his son Robert Winterscale who confirmed the gift in his will of 1723. The will states that it was for the poor of St Margaret's parish, who were to be chosen by the feoffees (trustees of charity land), and who would live in the six rooms rent free. It apparently incorporated remains of a fifteenth-century timber-framed building parallel to the street, which had been re-roofed in the seventeenth century. There was a malt kiln in the yard adjoining, and the rooms above those used by the almspeople were let as corn-chambers – and no doubt attracted vermin. In 1946 these upper floors were derelict and unsafe to walk on. Two houses, 93 and 95 Walmgate, had a back door onto the court, but were not for almspeople, even though they were linked to it and shared the lavatories in the yard.

William Hargrove reported in 1818 that three people were living in one room, apparently quite happily – an aged couple and their elderly daughter. By the time Bill Sessions visited it in 1946 it was already listed for demolition. Only three ground-floor rooms were occupied and it was in very poor condition and extremely damp. The floors were old brick (a perfect absorber of damp from the soil they were laid on) with linoleum here and there – which when lifted was slimy and black with the wet. Mattresses had to be dried before the fire. The ladies were evidently trying to make the best of what they had but one, Mrs Goldborough, who had been in residence for only eighteen months reported that three or four of the ladies had died in that time, mostly from coughs (or pneumonia?). Another, Mrs Cox, made grim jokes about death playing "puss-puss-come-to-my-corner" round the yard.

Winterscale's Hospital under demolition 1957
The name plate above the door reads "Winterscale's Hospitals"
The malt-kiln chimney can be seen above
(*City of York Libraries. Archives and Local History Dept*)

The *Yorkshire Gazette* (27 July 1951) reported a Housing Committee decision to take no action to restore the buildings for continued use as almshouses. The charity lacked resources for repairs and the almshouse was demolished in 1957.

(30) Wandesford's Hospital 37 Bootham

Mary Wandesford's will of 1725 made provision for the purchase of a building suitable for an almshouse. None was found and a site was bought from William Wilberforce of Hull just outside the city walls in Bootham where the present house was built in 1739. Payments were made to John Terry for carpentry, and to Robert Kibblewhite, Thomas Dunn and Richard Nestrop (Nelthorpe?) for brickwork. Some of these had previously worked on the Mansion House. The house is south-facing, set back from the street, a red-brick two-storey building with seven bays under giant arches, a projecting pediment over the three central bays, and a high, pitched roof. A bust of Mary Wandesford sits in a niche above the front door. The design has sometimes been attributed to John Carr, but as he was only sixteen when it was built, this is unlikely. Akehurst (2008) suggests the probable architect was James Moyser, a colleague of the Earl of Burlington who introduced the classical style to York. Ann Routh's almshouse (1749) and Tymperon House, both in Beverley, have very similar features, and were designed by Moyser.

Ann Routh's, 1749, Beverley

Mary Wandesford's, 1739

The front railings were removed in 1941 for the war effort, and no doubt dumped at sea like the rest. They were finally replaced in 1985/6. The back gardens were larger than those at the front including, in the eighteenth century, not only privies but an orchard which was let to raise some income. The orchard was sold in the 1950s and is now occupied by Bootham School's tennis courts. Residents make great use of both front and back gardens, though the latter has now (2010) been overshadowed by the prison-like back wall of a block of flats in Bootham Row which has taken both light and a unique view of the Minster. No opportunity was given to object to this development.

The house was for ten poor gentlewomen who were never married, and who practised the established religion, "who shall retire from the hurry and noise of the world into a Religious House of Protestant retirement ... and they shall be obliged to continue there for life". These requirements led to its being called for many years The Protestant Old Maids' Hospital. Mary Wandesford's will was contested by her family in the Court of Chancery in 1737, possibly because the family resented her money being left away from them, but equally possibly because the original terms of the will quoted above suggested that she wanted the almshouse to be a closed house for younger women – in effect a sort of nunnery. This smacked of popish practices and may be why the Court of Chancery in upholding the will decreed that the minimum age of residents should be fifty, which it remains. The first old maids took up residence in 1743.

One of its endowments was a farm near Brompton-on-Swale which is still called Old Maid's Farm, though it no longer forms part of the endowment. It was sold in 1951 when, according to the minutes of trustees' meetings, it stopped being a source of income and began to require money for repairs. The rest of the endowment came from Mistress Wandesford's own wealth, some of which was in South Sea stock. This paid for maintenance and stipends. In 1909 eight of the residents received yearly stipends of £26, the remaining two received £15 each. At that time the residents were largely daughters of clergy or professional men, or former ladies' companions and governesses.

Bill Sessions visited the house in 1946, when each flat had an upstairs bedroom, and a sitting room on the ground floor with a kitchenette off, with a grill only. If anyone wanted to cook using an oven they were supposed to use the Porteress's cooker but the arrangement had not proved satisfactory and had lapsed. Some had a range and oven in the sitting room. The residents shared washing and bathing facilities – one bath for the whole house "a relatively recent addition", and the gas water-heater would not function if a cold tap was in use anywhere else. There was electric light, supplied free by the trust, but the gas supply was separately metered. At this period the trustees' meetings were held less than once a year under the chairmanship of the Dean of York, Eric Milner White, and the living conditions of "inmates" were not much discussed. One independent-minded lady had paid to have a cold tap and sink, and a gas fire fitted in her flat so that she did not have to carry water or have coal stored in the flat.

The whole almshouse interior was renovated in 1968 to designs by architect Michael Mennim, at a cost of about £21,000, and carried out by William Birch & Sons. The money was raised through grants and a Corporation mortgage. Mary Wandesford's arms and a smaller pediment were both installed over the front door at this time. The arms were repainted and regilded in 2009. The 1968 refurbishment created twelve self-contained flats, each with its own kitchen and bathroom, and central heating. Four of the flats were made into bedsitting rooms, two on each floor; four into one-bedroom maisonettes with an internal staircase; and four into one-bedroom flats, two on each floor. A common room was created on the ground floor which provides a daily meeting place for anyone who wants company. Above there is a chapel, where an Anglican chaplain still takes communion once a month. Bathrooms and kitchens were upgraded in 2004 at a cost of £60,000 (paid for partly with an interest-free loan from the Almshouse Association).

The exterior of this grade 2* listed building has often needed attention. Part of the roof collapsed in 1927; it was finally repaired in the 1932 at a cost of £1,819 14s 2d. A flat roof covering the gulley between the two pitched roofs was installed during the 1968 refurbishment. Strengthened in the 1990s, it has in 2010 been refelted and partly retiled. Decayed brickwork along the whole front plinth was replaced in 2008.

The almshouse is still functioning, the only foundation of such long-standing which remains in its original building. Marital and religious restrictions have been lifted so that widows and divorcees may apply, and they need only profess the Christian faith, not just its Anglican form.

(31) Margaret Mason's Hospital 21 Colliergate

Margaret Mason was the widow of Thomas Mason, Sheriff of York in 1701. In her will of 1732 she left money to establish her almshouse in Colliergate, originally for six poor women. It was endowed with houses and land in Fossgate, from which each resident was to receive 20 shillings per annum. The endowment was sold in 1904 for £600. In 1909 the five residents received £5 a year each. The endowment produced an annual income of about £64 in 1939. Two other benefactors contributing to the maintenance of the almshouse and the support of its residents were a Mr Richardson of Fulford and Lady Conyngham, a benefactress of some fame in York.

The residents were poor women of good character, aged over sixty who had been rendered poor by misfortune.

Margaret Mason's former hospital, 21 Colliergate

Originally timber-framed, the house is a three-storey brick building of three periods, mostly from the late sixteenth to early seventeenth century, as well as earlier. The front and rear walls were rebuilt in brick in 1783 with money from a legacy. In the mid-nineteenth century the floors at the front were raised level with the back, and, at some point it was converted into five bedsitting rooms with a ground-floor laundry and drying room. A small opening called Beans Passage led off Colliergate to a dark entrance and the stairway to the almshouse flats. Bill Sessions reported in 1946 seeing signs of damp from a leaking roof. His guide Mr Holmes said re-roofing was desirable but funds did not allow. There was no bath, and the only lavatory was an old WC out in the yard – as Bill Sessions commented, "two flights down and back again for the top floor". The only light in the stairway came from a small roof light, which had been painted out during the war. One of the ladies "mildly" remarked, "it was a wonder none of us had an accident". There were three ancient gas jets on the stairs but two seemed to have been out of action for some time, and the one at the bottom caused problems about who was to turn it out "so it too was seldom used". The management seemed to have fallen into careless ways, not contacting the Charity Commission, for example, on the death of a previous clerk.

Bill Sessions was unimpressed by the state of affairs in the house and when his guide mentioned that he had hopes that the Nuffield Foundation would make grants available to re-roof and restore the property he remarked in his notes "But I did not encourage [these hopes] too much".

The almshouse closed in 1958. It is now a shop with a flat above.

(32) **Briggs' Charity** Water Lane

The almshouses were established by a deed of 1802 in Water Lane (probably First Water Lane, now King Street). It consisted of a house with a cellar, for two poor inhabitants of the parish of St Michael's Spurriergate and was administered by parish feoffees. The house was demolished for street improvements in 1853 and was exchanged for one in Low Ousegate, owned by T.J. Gaussen. This house may have been one listed in the York directory for 1858 as Sanderson & Gaussen's ironmongers business at 1 Low Ousegate. If so, the almspeople may have occupied an upper floor of the building. In 1868 the house was occupied by four aged people, but failed to feature in a survey of 1907. It was presumably defunct so Bill Sessions did not visit or report on it for the Nuffield survey of 1947. The charity was probably merged with the parish estate.

(33) **St Martin's Cottages** Coney Street

These chantry cottages had been in St Martin's courtyard, Coney Street, since the mid-eighteenth century and were mentioned by the Brougham Commission in 1824. They were not the subject of a specific trust, but had been held by successive feoffees for the benefit of the parish, and let by overseers of the poor, without rent, to the poor of the parish. This had effectively made them almshouses. They consisted of a three-storey brick building, occupied by six people, mostly poor widows or spinsters of the parish. Each had two rooms, and later paid a nominal annual rent of six shillings.

Bill Sessions missed these dwellings in his account of York almshouses for the Nuffield report. When he finally visited them in 1947 he found the cottages attractive. They were built at right-angles to Coney Street "down the alley leading past the burnt-out church of St Martin's to the Herald Printing offices". They were of brick and tile, with exterior woodwork tastefully and recently painted. Cosy rooms led off a small passage. They had low ceilings, an open range with oven, and electric light. There was a cold tap on each landing, shared between the two residents on each floor. "All slops had to be carried downstairs to the lavatory"; ashes from the fires had to be carried down too. Coal was kept in a cupboard in the living room. Only three women were living there, the top floor having been empty for some time "probably because the stairs had become too difficult." One of the residents, a Miss Parkinson, was evidently rather fierce and frightened the others. This was becoming more trying as the cottages were gradually being left empty so she had fewer neighbours to terrorise.

The only house remaining in the lane by St Martin's Church Coney Street

After the firebombing of the church in 1942, and despite the proximity to fire and firemens' hoses over the three following days, the ladies had stayed put. The editor of the *Yorkshire Evening Press*, Mr Cobham, commented on how difficult it had been to move the old folk at the time of the raids. By 1949 only two people were living there, and the cottages had become unfit for habitation. The property was sold that year and has since been demolished. A remaining house and shop, on the corner with Coney Street, show how attractive the row once was.

(34) St Catherine's 1835. Rebuilt in Holgate Road for four widows to live in. *See* **(6)**

(35) Lady Hewley's Originally 1700. Tanner Row. Rebuilt 1840, St Saviourgate. *See* **(26).**

(36) Ann Harrison's Almshouses, St John Street
Later in Abbot Street/Penley's Grove Street

The original hospital was built on the corner of St John Street and Penley's Grove Street in 1845. It was a high single-storey building consisting of eight tenements of brick and stone with a central chapel, standing in its own grounds surrounded by a low wall and railings. It was established by deeds of 1833 and 1839 for eight poor Anglican women over fifty. The chapel ceased to be used for services after 1900.

At selection, preference was given to those in reduced circumstances but with some means of their own. They received stipends from dividend income: in 1909 this amounted to £5 a quarter with an extra £1 at Christmas; in 1946 it was worth £2 per month.

A glass and wood porch (a post-1845 addition) led into each good-sized bed-sitting room. There was a kitchen with an open range and small cold-water sink, with a pantry next to the back door. Bill Sessions, visiting in 1946, found that the shortages of wartime had left it in poor condition. There was no money for repairs, maintenance or structural improvement, though the small stipends continued to be paid to the residents. The house still had no bath, and lighting was by gas. Each dwelling had a small yard surrounded by a high brick wall, in which were a coal house, an outside WC and an extra outhouse. One lady, slightly better off than the others, had made improvements including an up-to-date fireplace and modern range, and an Ascot water-heater (seldom used, however, as it tended to pop when lit which frightened her).

Parts of the chapel cornice had twice fallen through the roof of the next door resident, narrowly missing her. The garden was desolate, being overgrown, with small Anderson air-raid shelters awaiting demolition and the remains of a static war-time water tank. There was also a problem with local youth who threw stones at the windows, and once with a masked man who had burst in demanding food.

[Note by Bill Sessions, 2010] When I became a trustee in the 1950s plans were gathering pace for building replacement almshouses at the bottom of the gardens. Trustee Lewis Waddilove's work for the Joseph Rowntree Housing Trust gave him experience in raising funds for house-building and he produced a scheme which would just finance Harrison's proposed new building, from sale of all investments and part sale of land, albeit involving "dangerously" high levels of interest on borrowed money into the long-term future. Harrison trustees were nearly set to go forward on this basis when an anonymous benefactor, having heard of these plans, offered a substantial sum in memory of her late husband which would finance nearly half of the £12,000 cost for the new building. She wished to remain anonymous but her gift is naturally remembered with gratitude. It enabled the Trust to proceed without the burden of interest.

Eight new cottages were built in 1964 on the next corner, along Abbot Street, and the old almshouse was demolished. A copper tube containing everyday objects was buried near the foundation stone and the ceremony was recorded in the *Yorkshire Evening Press* of 23 July. A new Charity Commission scheme was drawn up in 1965, amended in 2003, offering accommodation to poor widows or spinsters, aged over fifty. The new accommodation consisted of a bed-sitting room with separate kitchen and bathroom. The cottages were built round three sides of a court, which contains an attractive garden. Underfloor heating was installed so there are no chimneys. This background heating is controlled by the clerk, and residents also have their own electric fires and meters.

Ann Harrison's Almshouses, Abbot Street
From the garden
(Penny Smith, Chair of Trustees)

York Corporation duly bought the old site and built, appropriately enough, a communal home for about forty people.

(37) Bishopthorpe Poorhouse/Almshouse
Main Street, Bishopthorpe

Bishopthorpe parish supported an eighteenth-century poorhouse for nearly two centuries. Built by the Overseers of the Poor "on the waste" (at the end of Sim Balk Lane), with two rooms for one occupant, it appeared in Archbishop Drummond's Visitation returns in 1764. When its last resident died in about 1938 it was pulled down. In the mid-nineteenth century, however, four other cottages were built nearby, also for the poor, on the corner of Main Street and Sim Balk Lane, opposite the Pinfold. It is not entirely clear what their status was: Camidge (1890) suggests they were founded as almshouses by Archbishop Vernon Harcourt, though the documentary evidence for this is lacking. However, both the original poorhouse and the newer buildings are marked on a map used for the Valuation Office Survey of 1910-1915, and are described generally as "poorhouse" in the Field book of the survey (National Archives IR58/94764). This also gives the names of occupants and the rents paid – which suggests they were not almshouses as such. Such rents would have been applied in aid of the parish poor rate. Both buildings were reported to be owned by the Overseers, and clearly the Parish Council managed them.

The nineteenth-century cottages consisted of two dwellings with two rooms each, and two with one room each, all designed for single occupancy. There was a central archway leading through to a yard, where there was a pump and "2 closets & ashpits for 5 houses". Maintained (after a fashion) by the Parish Council, they were several times reported as unfit for human habitation and repaired. They continued to be occupied by elderly widowed villagers until 1953. After this the cottages were let to local businesses (Mr Gonsalves, a watchmaker had a workshop there in 1954) and finally sold to building contractors, Messrs Bradley. Part of the land was also sold to the West Riding County Council for the widening of Sim Balk Lane at its junction with Main Street. The cottages were finally demolished in 1963. There is a photograph of them in *Bishopthorpe remembered* (Cantrell, 1988).

(38) Barstow's Hospital Moved from Blossom Street in 1859 to the corner of Caroline Street and Clement Street at the Clementhorpe end of Nunnery Lane.
See (**25**) and also (**46**).

(39) St Thomas's Hospital rebuilt 1863 next to Victoria Bar, Nunnery Lane. *See* **(9)**

(40) Sister Wilson's 17-19 East Parade, Heworth

These almshouses were founded by Ann and Mary Wilson in 1885 at 17-19 East Parade, Heworth. The sisters were milliners in High Ousegate, who retired to Torquay. The almshouses were for ten poor women over sixty who had lived in York. They consisted of two brick and slate houses on the street, and two cottages behind. These houses and the cottages behind appear on the 1852 map at the further end of the first row of houses in the street from the Mill Lane end.

Supporting income came from property in York and Middlesbrough, and Skeldergate Bridge tolls. It was administered by York Charity Trustees, the clerk being Charles Robinson who was responsible for several other almshouses.

When Bill Sessions visited, there were three residents in number 17, and two in number 19; only one of the cottages was occupied. The houses had two bedsitting rooms downstairs and two upstairs, and took four residents each; the cottages took one each. The trustees had wanted to adopt a two-room system, but this had been rejected – the newer residents said it would be too much to keep clean and they had too little furniture to fill them. One house had an attic space available for use by a resident. Each room had a fireplace, oven, and gas lighting. The stairs in the houses were lit by a single gas lamp at the foot – installed after "years of application and delay", and still they were dark. Outside, in each long narrow garden, was a washhouse with sink and cold water tap – the only tap for the four residents of each house. Next to it was the shared WC, coal store and a small surface air-raid shelter. The washhouse at number 19 was attached to one of the ground-floor flats whose occupant permitted the other three to get water, but apparently not to wash clothes in it – they did not dare to, at least. Nothing had come of their requests for sinks and taps in their rooms.

Down the garden were the two cottages each with a living room downstairs, one with two small bedrooms upstairs, the other with one large bedroom. Each had an indoor sink and cold tap at the foot of the stairs, but the two dwellings had had to share an outside WC, though in 1946 there was no-one to share with as one cottage had been empty for some time. They faced onto a lane at the back, parallel to East Parade, with a door and passage into the common garden.

The front in East Parade had a thin strip of garden with a low wall – the railings and gate of one removed for the war effort. The back garden was untended and overgrown. The houses and outhouses behind were demolished in about 1957. The site is now a carpark.

The Charity was amalgamated with forty-five other endowed municipal charities in 1956, and is administered as part of York City Charities.

(41) **Ellen Wilson's** Lawrence Street

Ellen Wilson's Hospital is next to St Lawrence's church. It was founded in 1894 in memory of his mother by Dr John Sykes, a physician in Doncaster who had been born in the parish of St Lawrence. On the death of Dr Sykes on 1 April 1901, his residuary estate passed to the charity, much of it consisting of investments in railway stock.

Ellen Wilson's Hospital, Lawrence Street

The architect, A.S. Ellis of Westminster, built six cottages in three pairs, in red brick with stone facings and stone-mullioned windows – seventeenth-century style.

They face a garden and sit at right-angles to Lawrence Street. Each pair has its own main door and entrance hall. There is a shaped pediment over the entrance and a cupola on the roof with an ogee cap. The almshouses were originally intended for aged or infirm women born or resident in St. Lawrence's parish, selected without religious test. Latterly it has been for the benefit of Anglican widows and spinsters of the parish.

When Bill Sessions saw it in 1946 the gardens looked pleasant, and included a vegetable garden which was separately let. Each cottage consisted of a living room with bed recess screened off by a curtain. Behind was a small kitchen with cold tap and sink. Two or three residents had gas cookers. Lighting was by gas, but electric light was being installed. There was an indoor WC, and outside coal store. There were no baths. The trustees at that time took a keener interest than had been observed in some other almshouses and seemed to be on friendly terms with the ladies. The residents were receiving 8 shillings a week.

In the 1980s the cottages had become very down-at-heel and were facing a closure order. However, they were refurbished in 1994, with help from the City Council, but are now threatened with being surrounded by student housing. The charity is also involved in a possible merger with Dorothy Wilson's **(29)**.

(42) Terry Memorial Homes, Skeldergate

Anne Middleton's Hospital, and Terry Memorial Homes, Skeldergate.
Now a hotel

Sir Joseph Terry (1828-1898) was the son of the founder of Terry's York confectionery business. He served both as Sheriff and Lord Mayor of York. These two almshouse cottages were built in memory of him to designs by Penty & Penty in 1899. They stand in front of Anne Middleton's Hospital **(20)** in Skeldergate at right-angles to the road with a tablet set into the end wall facing the street, which reads "These almshouses were erected by public subscription to the memory of the late Sir Joseph Terry 1899". The subscription raised £1,020 12s 9d. The cottages were for two married couples over sixty. In 1946 one couple was Mr and Mrs Leeman. Mr Leeman had been a railwayman who had been paid to look after the gardens before retiring. He continued to do the garden. There was no stipend. The ladies at Middleton's Hospital told Bill Sessions that this "likeable rough talkative fellow" was kept pretty closely in hand by his wife.

Tablet recording the foundation of Terry Memorial Homes, Skeldergate, with his arms
"These almshouses were erected by public subscription to the memory of the late Sir Joseph Terry. 1899"

Each cottage had a living room with a bed recess screened off by a curtain ("only just large enough for the double bed"). There was an old-style fireplace with an oven; behind it a scullery with sink and cold water tap, and a small gas cooker. The lighting was also gas. Outside the back door was a coal store and WC. During the 1942 air raids a bomb had fallen between the cottages and the river (a very short distance). Several other houses had been destroyed, but they had suffered relatively minor damage. The almshouse closed in the 1970s, and now forms part of Lady Anne Middleton's Hotel. Like Anne Middleton's, it was a municipal charity which had been managed by York Charity Trustees until 1956 when both were amalgamated with forty-four others to be administered as York City Charities.

(43) Lady Deramore's Memorial Almshouses
The Village, Heslington

Robert Wilfred de Yarburgh-Bateson, third Baron Deramore, founded these four almshouses in 1903. A large memorial tablet in the central pediment reads "These almshouses were erected by Robert Lord Deramore in memory of his beloved wife Lucy 1903". Lucy, Lady Deramore, whom he married in 1897, died in 1901.

Lady Deramore's Memorial Almshouses, The Village, Heslington

They are built of brick with pantile roof. The architect was probably W.H. Brierley. The cottages were intended for spinsters or widows of the parish. Lord Deramore endowed the foundation with £1,125 stock, the income from which would provide the inmates with a pension of £10 per year. It is now managed under the Heslington Almshouses Charity together with Sir Thomas Hesketh's (**16**).

(44) Jackson Barstow's Almshouses
33-39 Front Street, Acomb

J.J. Jackson Barstow's former almshouses
Front Street, Acomb

The Jackson Barstow almshouses were built in Front Street, Acomb in 1910, and intended for widows or spinsters. A plaque on the front contains the Jackson Barstow motto "Nil Desperando Auspice Deo. Erected by J.J. Jackson Barstow 1910". Having missed them in his 1946 survey, Bill Sessions visited in 1949 and spoke to two residents, Mrs Terry who told him she received 10 shillings a year, and Miss Marley who told him there was electric light and a gas cooker.

In 1892, according to Bulmer's *History and directory of East Yorkshire,* John Jeremiah Jackson Barstow was Lord of the Manor of Storwood, near Wheldrake. His name is also associated with several modern care homes and hospitals in Somerset where he lived later. He died in 1939.

By 1963, as a result of the falling value of the pound and its effect on the endowment, the almshouses were let at a nominal rent to bring in some income. They have since been converted into four houses and have ceased to be almshouses.

(45) Colton's Hospital. Moved 1910 from Tanner Row to Shipton St, off Burton Stone Lane. *See* **(28)**.

(46) Cameron Walker Homes Bishopthorpe Road (also known as Walker Barstow Homes and Cameron Walker Court)

Charles Cameron Walker was a member of the Cameron brewing family. He died in 1907, and his almshouses were established in Bishopthorpe Road in 1909, becoming a registered charity in 1912. The endowment was £6,250.

Walker Barstow Homes, Bishopthorpe Road

Designed by W.H. Brierley, the main house consists of twelve three-room dwellings for poor men and women living within ten miles of York. It has three doorways, each opening onto a hall with two flats on each of the two floors. The flats as built contained a living room with range and oven, and a separate bedroom. There was a narrow scullery with sink and cold water tap and a small gas ring and grill. The wash-house was in the garden and contained two coppers, two sinks, cold water taps and mangles. Most of the residents preferred to do their washing inside in 1946, when Bill Sessions called. Each pair of flats shared a WC, but there were no baths installed when the house was built. Electric light, installed from the beginning, and water were provided free. Bombs fell nearby in 1942, but it was only when a British bomber crashed a short distance away in 1945 that any damage was done.

The twelve original residents included two couples (one of whom was the caretaker and spouse). The others were widows or spinsters, each receiving a small pension of 5s 6d per week in 1946. Couples received about 8s per week pension.

Barstow's Hospital (**25** and **38**) was demolished in the 1960s, and in 1972 the charity was merged with Cameron Walker Homes to become Walker Barstow Homes. A warden's bungalow and two others for residents were added at this time. In 1980, 1 Aldreth Grove was bought and converted into two flats.

Modernisation of the main block was carried out in 1982, which has added a bathroom to each dwelling. A communal lounge, known as the Barstow Lounge, was added in 1993, and in 1996 a house in Cameron Grove was purchased and converted into two more dwellings. The almshouse now houses eighteen residents including a caretaker. It became a registered social landlord and housing association on registration with the Housing Corporation, but intends to deregister when it becomes possible.

(47) **Burrill Homes** Water End, Clifton

John Burrill was a wealthy grocer, a partner of Borders, formerly in Coney Street. He was frequently asked to judge hams round the local fairs. He was a canny judge: every year he kept finding one enormous ham which regularly took first prize until one year he took a silver skewer along. Plunged into the giant ham it released a putrid smell – "End of prizes for it". Bill Sessions heard this tale from Mr K. Mackay of the Red House, Duncombe Place, and recorded it separately in his notes in 2007.

Burrill died in 1924 and the executors of his will established the almshouses on Water End, designed by architects Ward and Leckenby and built by William Birch & Sons in 1931. The dwellings and site cost £10,631 according to Bill Sessions' notes in 1946. Their attractive design and layout is modelled on a seventeenth-century almshouse style. They are set back from and above the road, behind wrought-iron railings. These very fine railings and gate were saved from removal and destruction for the war effort (significant pressure was exerted to prevent it in 1944) and are under restoration in 2010.

The almshouses consist of six one-storey dwellings and a central house, set in three ranges round a formal garden, and were originally intended for members of the grocery trade and their families. A weekly stipend of ten shillings was paid to each resident, and they received free electric light, and water.

The almshouses face south and each has a bay window, above which small carvings have been set into the stonework: some are of animals, some of faces. Each bungalow has a sitting room and bedroom, kitchen and bathroom. In 1946 there was an open fire; the kitchen had a gas cooker, and a sink with both hot and cold water – the first York almshouse, perhaps, to achieve such luxury – and the bathroom contained sink and WC (and by implication also a bath). Bombs fell nearby, in the area of the present youth hostel, in 1942, which badly frightened the residents but did no damage.

Burrill Homes, Water End, Clifton

Each cottage also has a small garden area of its own and at the back of the site are allotments and an orchard for the residents. In 1946 the trustees were disappointed by how little the front garden was used. The residents preferred pottering in their own back gardens as, apparently, they still do.

The residents are men and women over sixty who are preferably church-goers, though all applications are considered. The central two-bedroomed house is occupied by the caretakers, normally a retired couple. Stipends are no longer paid, but for many years no charge was made for accommodation. Since 2007, however, residents have been asked to make a small contribution towards heating costs.

(48) **Fothergil Homes** Avenue Road, Clifton

These homes, on the corner of Avenue Road and Avenue Terrace, Clifton, were built in 1935 in memory of Colonel Thomas Fothergil by the bequest of his niece, Mrs Jessie Ashton, who had inherited her wealth from him. He was a nineteenth-century Yorkshire businessman and MP. The Charity Commission scheme of 1936 fixed the number of almspeople at ten "poor working people of good character, ten years resident in York": five men and five women, each to receive stipends of not less than 5s per week, a figure which had doubled by 1946. The Homes are now administered by York City Charities.

Fothergil Homes, Avenue Road. Original 1935 cottages

Designed by Ward and Leckenby, the ten brick and tile almshouses were constructed at right-angles round a broad square of garden. Each has an entrance hall, separate bedroom and living room with, in 1946, an open fire. The kitchen had a gas cooker, and sink with cold and hot water – heated from the back of the fire. There was cupboard space outside for tools, but little cupboard space inside. Each had a lavatory, washbasin and bath. Bill Sessions remarked in his notes "we heard that the baths were seldom used" – which might have been prejudice on the part of the teller, who may have referred mainly to the men. He talked at length to Mrs Rougier at no. 11, and to the warden at no. 7, Mrs Dudding, whose duties were limited to notifying the trustees of repairs, and relatives in the case of illness. She would help the ladies, but the old men were too much: "some of [them] were dirty and did not manage very well unless a relative came in with a meal for them".

A bomb fell in 1942 completely destroying the entrance gate and rendering the homes uninhabitable for a time. Tiles, windows, plaster, and doors were damaged but no one was seriously hurt – just terrified. They removed "to relatives etc" while first aid repairs were carried out.

After the war, there were developments which have made Fothergil Homes the largest group of almshouses in York, providing accommodation for up to sixty people, including couples. This has been the result of the amalgamation of the forty-six municipal York charities, including several almshouse charities. When Ingram's Hospital (**19**) closed in the late 1950s, land behind Burton Croft, John Bowes Morrell's house in Burton Stone Lane (demolished 2005), was transferred to Fothergil Homes to house the displaced residents. The dwellings built for them form a very traditional almshouse-range closing off the space on that side. With money from the sale of other municipally-owned almshouses, such as Middleton's, Terry's, St Thomas's, etc., further additions were made in the 1970s. In 2003, National Lottery funding enabled the erection of an attractive day-room for use by the residents and one or two societies in the area. It is attached to the offices and community room and is well used. The gardens too are well used: some residents grow vegetables as well as flowers. The central open space, though large, is private enough to be a welcoming space and as they have to cross it to reach the community rooms, the residents are able to meet there to chat.

After recent fundraising success, there has been a good deal of upgrading, including the installation of central heating. An ongoing maintenance programme is based on a twenty-year cycle of four tranches of five-years' work to bring the older property in particular up to date. Recent maintenance has included replacing metal window frames as well as boilers. As dwellings become vacant they are assessed for potential upgrading, in particular for new bathrooms and kitchens The residents' contributions cover the cost of all work, and include all heating, lighting, and water.

There is one full-time, and one part-time warden, and a gardener and handyman, who between them manage the property, the garden and the residents.

Fothergil Homes, Avenue Road

(49) **Sir John Hunt Homes** Fulford

These very distinctive and attractive two-storey dwellings with their shaped gables were founded in the mid 1950s by Sir John Hunt, a local brewery owner. They were built in red brick and tile to a design by architects Needham, Thorpe and White, and grouped round a large green space on the corner of Fulford Road and St Oswald's Road, with smaller blocks of dwellings behind. (Not to be confused with Connaught Court, the neighbouring residential care home).

Sir John Hunt Homes, Fulford

There are two houses for married couples, and twenty-four cottages for single people who are in need, and aged between fifty-five and seventy at entry. As at the Burrill Homes on Water End, the front gardens are not much frequented by the residents, who may well feel rather exposed in them.

The present charity was created by the trustees of Sir John's nephew Reginald Hunt in 1978, at which time the residents were asked to pay a maintenance contribution charge of £5 per week. They had previously lived there free of charge and had received a stipend or pension from the trust of between £1.50 and £2.50 per month. The income from the original endowment had ceased to produce enough to maintain what is a large property. There was a certain amount of soul-searching over this, but the charge was fixed after consultation with the Rent Officer and there was potential assistance from the Department of Health and Social Security, for people unable to pay. Now at more than fifty years old, the almshouses are carrying out a maintenance programme which has so far included the roofs, chimneys, windows and doors.

(50) Jane Wright's Charity

5-9 Ogleforth and 1-5 Saville Court.

The charity was founded in 1675 by Mistress Jane Wright, who left money for the relief of poor widows and housekeepers, for apprenticing poor boys of the area, and for setting up apprentices in trade. Money was invested in land and farms to bring in an income for the purpose. By 1934, however, there was no one eligible for the apprenticeships, and the whole was spent on supporting the poor.

In the 1960s some pensioners of the charity were living in number 9 Ogleforth, at that time a semi-derelict nineteenth-century house, itself a reconstruction of an earlier building. The then Clerk arranged for it to be bought; he sold it without profit to the charity a few years later and a conversion into four flats was carried out by Michael Mennim. Each flat contains a bedsitting room with separate bathroom and kitchen. An original plan to have a shared kitchen-cum-meeting place was discouraged by the General Director of the Almshouse Association as he thought the cooking equipment was more likely to be used for purposes of personal violence – one can only assume he had special knowledge of 1960s Ogleforth. The conversion was reported in the *Almshouses Gazette* of September 1972 which noted with particular interest that this was a charity whose original trusts did not include the provision of housing, and that by purchase and restoration it had saved for posterity a building whose disappearance (demolition was guaranteed at that date, despite its location) would have been a grave loss. It may be noted that Mrs Phyllis Richardson, the wife of the Dean, had been partly responsible for this restoration, and with Michael Mennim had also modernised Wandesford House a few years earlier. The Dean's wife is often a trustee of both almshouses.

Jane Wright's Almshouses, Ogleforth

The charity then sold the farms at Rufforth and Poppleton, and bought property on the Mount, rent from which enabled the purchase of the neighbouring houses, 5 and 7 Ogleforth, which date from c1830. This purchase included the yard and other buildings behind, where a Victorian carriage-maker had once had a workshop. With the help of a Housing Corporation loan, work was started in 1981 on a £180,000 project to convert the old houses and build new ones round the internal courtyard, now known as John Saville Court. The architects were Ferrey and Mennim, the builders William Birch & Sons. There are now eleven dwellings for almspeople, including the caretaker's house at no. 7. The flats vary in size and shape, but most provide a bedroom, sitting room, bathroom and kitchen, and they are centrally heated. The new flats in the court are light and airy. The ground-floor flat in no. 5, now rather darkened by a new building opposite in Ogleforth, still has an old range *in situ* in the kitchen (as a feature rather than a functioning installation).

The naming of Saville Court maintains a link with the former chemists on the corner of Aldwark and Goodramgate, often referred to in the past (and sometimes still) as Saville's Corner. The Saville family still keeps a link with the charity.

The gates to Saville Court, Ogleforth

Table 1: Almshouses in York 1947-2010

Listed in the order given in the Nuffield report, *Old people*

Almshouse	Type	No. of residents	Change since 1947
Thomas Fothergil	Municipal	10 in 1935. 60 in 2010	Sixfold increase in numbers
Ingram's Hospital	Municipal	11	Bought 1957. Five flats for rent 1959.
Middleton's Hospital	Municipal	12	Hotel
St Catherine's Hospital	Municipal	4	Demolished
St Thomas's Hospital	Municipal	12	Hotel
Terry Memorial Homes	Municipal	4	Part of Middleton's Hotel
Sir H. Thompson's Hospital	Municipal	6	Demolished c1949
Sir Robert Watter's Hospital	Municipal	7	Demolished 1956
Sister Wilson's Home	Municipal	10	Demolished c1957
Barstow's Hospital	Independent	4	Demolished 1960s: merged with Cameron Walker
Burrill Homes	Independent	6 (+ caretaker)	
Colton's Hospital	Housing Association	12	
Ann Harrison's Hospital	Independent	8	Demolished and rebuilt in 1964
Lady Hewley's	Independent	9 (+caretaker/superviser)	Bathrooms and kitchens added 1957, replaced 2010
Margaret Mason's	Independent]	5	Closed 1958. Shop
Cameron Walker	Independent	11	Merged with Barstow's
Wandesford Hospital	Independent	10 (incl. caretaker)	Converted into 12 self-contained flats
Dorothy Wilson's Hospital		16	Merger with Ellen Wilson's due 2010?
Ellen Wilson's Hospital	Independent	6	"
Winterscale's Hospital		6	Demolished c1957

Table 2
Almshouses in York in 2010 (including Heslington)
Listed in alphabetical order of benefactor's surname

Almshouse (map no) (Charity no)	**Type**	**No. of residents**	**Location**
Burrill Homes (**47**) (239036)	Independent	6 mixed (+caretaker)	Water End, Clifton
Colton's Hospital (**45**) (221281)	RSL and Housing Association	12 women	Shipton Street
Lady Deramore's (**43**) (224416)	Heslington Almshouses Charity: Independent	4 women	Heslington
Fothergil Homes (**48**) (224227)	York City Charities	60 mixed	Avenue Road, Clifton
Ann Harrison's (**36**) (221280)	Independent	8 women	Penley's Grove Street
Sir Thomas Hesketh's (**16**) (224416)	Heslington Almshouses Charity: Independent	6 mixed	Heslington
Lady Hewley's (**35**) (230043)	Independent	10 mixed (incl. caretaker/superviser)	St Saviourgate
Sir John Hunt Homes (**49**) (214489)	Independent	24 mixed	Fulford
Walker Barstow Homes (**46**) (222564)	RSL & Housing Association	18 mixed	Bishopthorpe Road
Wandesford House (**30**) (251379)	Independent	12 women (incl. caretaker)	Bootham
Ellen Wilson's Hospital (**41**) (232135)	Independent	6 women	St Lawrence Street
Dorothy Wilson's Hospital (**27**) (224417)	Independent		Temporarily lacking an almshouse
Jane Wright's Charity (**50**) (228961)	Independent	11 mixed (+caretaker)	Ogleforth Saville Court

References and booklist

This study is chiefly based on secondary sources, most of which are listed in the sections below. Many of the texts are also available online (as noted). Aspects of the subject such as the religious life of hospitals and almshouses, patronage, medicine, and architecture, are covered in more detail by other authors. The references below may provide the starting point for pursuing an interest in those aspects.

Newspaper cuttings form part of the documentation on York almshouses kept by William Sessions since 1949, and are listed here.

Social history: poverty, charitable giving, the poor laws

Atkinson, A.B. et al. *Parents and children: incomes in two generations*. London: Heinemann Educational, 1983 [a follow-up to the Rowntree reports on poverty in York]

Barry, Jonathan and Jones, Colin, eds, *Medicine and charity before the welfare state*. London: Routledge, 1991

Booth, Charles *Life and labour of the people in London.* London: Macmillan, [3 editions, 1889-1891, 1892-1897, 1902-1904]

Charlesworth, Lorie "The Acts of 1601: connections between poor relief and charity in a legal and local context" paper given at Voluntary Action History Society, March 2003 [online text]

Charlesworth, Lorie "Why is it a crime to be poor?" in *Liverpool Law Review, 21(nos 2-3) 1999*

Critchlow, Donald T. and Parker, Charles H. *With us always: a history of private charity and public welfare.* Lanham; Oxford: Rowman & Littlefield, 1998

Davidson, Clifford "Suffering and the York plays" in *Philological Quarterly,* 81, 2002

Gladstone, David, et al. eds. *Before Beveridge.* London: Civitas, 1999

Hargreaves, Peter K. "The ticking time bomb" in *SIPP Times* IT78 November 2010

House of Commons Select Committee on Social Security *Pensioner poverty: seventh report*, 2000

Huby, M. et al. *A study of town life: living standards in the city of York 100 years after Rowntree*. York: Social Policy Research Unit, 1998

Kay, A.C. & Toynbee, H.V. *Report to the Royal Commission on the Poor Laws and Relief of Distress on endowed and voluntary charities in certain places, and the administrative relations of charity and the poor law*. (Cd. 4593) London: H.M.S.O., 1909

Laycock, Thomas *Report on the state of York …* [appendix to the First Report of the Commissioners of Inquiry into the State of Large Towns and Populous Districts] York, 1844

Malpass, Peter *Housing associations and housing policy: a historical perspective*. Basingstoke: Macmillan, 2000

Mayhew, Henry *London labour and the London poor.* 4 vols. London, 1861

O'Halloran, Kerry, et al. *Charity law & social policy.* (Springer ebooks). Dordrecht: Springer, 2008 [online text]

Oxley, Geoffrey W. *Poor relief in England and Wales 1601-1834*. Newton Abbot: David & Charles, 1974
Razzell, Peter & Spence, Christine "Social capital and the history of mortality in Britain" [Letter to the editor of] *International Journal of Epidemiology* 34(2) 2005: 477-478 [online text]
Rowntree, B. Seebohm *Poverty: a study of town life*. London: Macmillan, 1901
Rowntree, B. Seebohm *Poverty and progress: a second social survey of York*. London: Longmans, Green, 1941
Rowntree, B. Seebohm *Poverty and the welfare state: a third social survey of York dealing only with economic questions*. London: Longmans, Green, [1951]
Salter, Tony, et al. *100 years of state pension: learning from the past*. Oxford [?] Faculty of Actuaries: Institute of Actuaries, c2009
Tierney, Brian *Medieval poor law*. Berkeley: University of California Press, 1959
Titmuss, Richard M. *Essays on the welfare state*. 2nd ed. London: Allen & Unwin, 1963
Walker, Alan. "Towards a political economy of old age" in *Ageing and Society*, 1, 1981, pp 73-94

History of almshouses

Akehurst, Ann-Marie "Wandesford Hospital: York, Colonel Moyser and the Yorkshire Burlington Group" in *Architectural History*, 51, 2008: 111-135
Archaeological evaluation: *Britannia car park, Heworth Green, York*. FAS (Field Archaeology Specialists Ltd), 2001 [Unpublished archaeological report]
Ayliffe, William "St Bartholomew's Hospital and the origin of London hospitals", [Transcript of a talk given at Gresham College, 7 April 2008; available online]
Aylmer, G.E. & Cant, Reginald, eds *History of York Minster*. Oxford: Clarendon Press, 1977 (1979 reprint)
Bailey, Brian *Almshouses*. London: Hale, 1988
Baines, Edward *History, directory, and gazetteer of the county of York*. Leeds 1822-23 [available online]
Brayley, C.E.W. *The annals of Bishopthorpe, 1215-1963*. Bishopthorpe: [The Author], 1964
Brunskill, Elizabeth *Some York almshouses*. (Occasional paper, 7). York: York Georgian Society, 1960
Buttery, Darrell *The vanished buildings of York*. York: [The Author], 1984
Camidge, William *Ouse Bridge to Naburn Lock* York: Reprinted from the *Yorkshire Gazette*, 1890
[**Cantrell**¸ Richard] *Bishopthorpe remembered*. Bishopthorpe: Archbishop of York's C.E. Primary School, 1988
Catholic encyclopedia: entries under *Alms and almsgiving*, and, *Hospital* [available online]
Clay, Rotha M. *The mediaeval hospitals of England*. London: Methuen, 1909 [available online]
Cullum, P.H. *Cremetts and corrodies: care of the poor and sick at St Leonard's Hospital, York, in the Middle Ages*. (Borthwick paper, 79) York: University of York, 1991

Cullum, P.H. *Hospitals and charitable provision in medieval Yorkshire, 936-1547.* York DPhil thesis 1989 [available online]
Dakin, Andrea Yarnell *A quiet corner of Skeldergate*. York: the Author, 1998
Dean, Gareth *Medieval York*. Stroud: The History Press, 2008
Dobson, R.B. & Smith, David M. eds *The Merchant Taylors of York : a history of the craft and company from the fourteenth to the twentieth century.* (Borthwick texts and studies, 33). York: Borthwick Publications, 2006
Dockray, K.R. "The Yorkshire rebellions of 1469" in *The Ricardian,* 6 (82) Dec 1983 [available online]
Drake, Francis *Eboracum: or, The history and antiquities of the city of York*... York, 1788
Dugdale, William *Monasticon anglicanum*. London, 1693 [available online]
Farrer, William ed. [Full text of] *Early Yorkshire charters* ... Edinburgh: [printed for the Editor], 1914 [available online]
Feinstein, C.H. *York 1831-1981*. York: Wm Sessions, The Ebor Press, 1981
Gaimster, M. and Bradley, J. "Medieval Britain and Ireland 2000" in *Medieval Archaeology*, XLV, 2001
Gilbert, Christopher G. "Ingram's almshouse, York" in *York Historian,* 1, 1976
Godfrey, Walter H. *The English almshouse; with some account of its predecessor the medieval hospital.* London: Faber, 1955
Granshaw, Lindsay and Porter, Roy, eds *The hospital in history*. London: Routledge, 1989
Green, Monica H. "Books as a source of medical education for women in the Middle Ages" in *Dynamis: Acta Hisp. Med. Sci. Hist. Illus.* 20, 2000, 331-369 [online text]
Hargrove, William *History and description of the ancient city of York*. York: Wm Alexander, 1818 [available online]
Hargrove, William Wallace *The history of York, its men & its institutions.* New ser. Section CXXXII: *Hospitals, almshouses and ancient chapels*. Yorkshire Herald Saturday supplement, 1907.
Hobson, John Morrison *Some early and later houses of pity*. London: Routledge, 1926
Hodgson, Geoffrey *A history of Acomb (Richardson's History, revised and enlarged)*. Rufforth: G.H. Hodgson, 2001.
Hodgson, Geoffrey "Wheels at St Catherine's" in *YAYAS Times* (York Architectural and York Archaeological Society) 1998, 15-20
Horden, Peregrine "A discipline of relevance: the historiography of the later medieval hospital," in *Social history of medicine, 1,* 1988
Howson, Brian *Almshouses: a social and architectural history*. Stroud: History Press, 2008
Howson, Brian *Houses of noble poverty: a history of the English almshouse*. Sunbury: Bellevue Books, 1993
Knight, Charles Brunton *A history of the city of York*. York: Herald Printing Works, 1944.
Margetts, F.C. *Records of the York Cordwainers Company from circa 1395*. York: Wm Sessions, the Ebor Press, 1983
Murray, Hugh, et al. *York through the eyes of the artist.* York: York City Council, 1990.

Norton, Christopher "The York fire of 1137: conflagration or consecration?" in *Northern History*, 34, 1998: 194-204
[Nuffield Foundation] *Old people: report of a survey committee on the problems of ageing and the care of old people* / under the chairmanship of B. Seebohm Rowntree. London: Oxford University Press, 1947
Orme, Nicholas & Webster, Margaret *The English hospital 1070-1570*. New Haven: Yale University Press, 1995
Pedersen, S. "Piety and charity in the painted glass of late medieval York" in *Northern history*, 36 (1) 2000, 33-42
Pevsner, Niklaus & Neave, David *Yorkshire: York and the East Riding*. (Buildings of England). 2nd ed. London: Penguin, 1995.
Potts, Richard A. *Dame Sarah's legacy*. Chester: Lady Hewley Trust, 2005
Rawcliffe, Carole "The hospitals of later medieval London" in *Medical History*, 28, 1984, 1-21.
Rawcliffe, Carole *Leprosy in medieval England*. Woodbridge: Boydell Press, 2006
RCHM (Royal Commission on Historical Monuments)
An inventory of the historical monuments in the city of York.
v.3: South-west of the Ouse. London: HMSO, 1972
v.4: Outside the city walls east of the Ouse. London: HMSO, 1975
v.5: The central area. London: RCHM, 1981
Report *of His Majesty's Commissioners concerning Lady Hewley's Charity... intended for the support of an alms-house in York...* Manchester: for the Ministers and Representatives of the Independent Churches of the County of Lancaster, 1829
Richards, J.D. *Union Terrace: excavations in the Horsefair*. (Archaeology of York, v. 11: The medieval defences and suburbs, fasc. 1). York: York Archaeological Trust, 1989
Richardson, Harold *History of Acomb*. York: Yorkshire Philosophical Society, 1963
Spall, C.A. and Toop, N.J. eds *Blue Bridge Lane & Fishergate House, York. Report on excavations: July 2000 to July 2002*. APC monograph series 001, 2005 [online text]
Standards of almshouse management: *a guidance manual for almshouse charities*. 5th ed. Wokingham: Almshouse Association, 2008
Sweetinburgh, Sheila *The role of the hospital in medieval England: gift-giving and the spiritual economy*. Dublin: Four Courts Press, 2004
VCH [Available online. Much of the relevant historical detail in VCH comes from Charity Commission files]
A history of the county of Hampshire, vol. 2 / H. Arthur Doubleday, William Page, editors. London: Oxford University Press, 1973
The Victoria history of the county of York / edited by William Page. London: Constable, 1974. v. 3 (pp. 336-352, by T.M. Fallow, 1913)
A history of Yorkshire: the city of York (Victoria history of the county of York) / edited by P.M. Tillott. London: Oxford University Press, 1961
A history of Yorkshire: the North Riding (Victoria history of the county of York) / edited by William Page. London: Constable; London: St Catherine Press, 1914-23
A history of the county of York: East Riding. v.3 / edited by K J Allison, et al. London: Oxford University Press, 1976
Wenham, Peter *The great and close siege of York, 1644*. Kineton: The Roundwood Press, 1970

York newspaper cuttings

York Herald articles
20 May 1837: *Poor Law Commissioner at York* (p.3)
Report of the great West Riding protest meeting (p.4)

Yorkshire Gazette articles
7 October 1949: *Almshouse site* (demolition order on Sir Henry Thompson's)
14 October 1955: *Haughton School to be closed* (reference to Lady Hewley's town residence)
30 August 1968: *Flats in old almshouse have been modernised* (Wandesford House)

Yorkshire Evening Press articles:
11 May 1949: Picture and article on Ingram's Hospital
18 May 1949: *Ingram's Hospital Norman archway* (Ingram's)
28 June 1957: *Another York "treasure" to be preserved* (Ingram's)
8 February 1962: Short item on Merchant Taylors', mentioning conversion of almshouse to cottage
2 October 1968: *Every 'mod. con.' now at York almshouse* (Wandesford House)
5 October 1972: *Corporation to buy 1863 hospital?* (St Thomas's)
19 April 1973: *£15,500 paid for York property* (St Thomas's)
16 June 1973: *1700 almshouse to put up tourists* (Middleton's)
2 March 1978: *Rising costs knock hole in Sir John's endowment* (Sir John Hunt Homes)
2 August 1980: *Benefactors who made sure that charity began at home* (Weekend Topic. Almshouses)
9 August 1980: *Charitable dame's generosity helped the poor* (Weekend Topic. Lady Hewley's)
8 July 1982: *Facelift for almshouses* (Colton's)
8 November 1983: *Duchess is to open York barracks* (Jane Wright's. Article)
25 November 1983: *In the alms of Jane: Duchess to see benefits of 17th century endowment* (Jane Wright's)
28 November 1983: *The Duchess of Kent unveils a plaque…* (Jane Wright's. Photograph)
18 July 1985: *Campaign to save old cottage* (White Row, 46 Front Street, Acomb)
26 September 1986: *Almshouses in York win award for conservation* (Jane Wright's)
14 August 1987: *Door closed* (Ingram's)
19 August 1993: *Foss almshouse to close down* (In Brief article. Dorothy Wilson's)
24 March 1997: *Who was Clifton's wealthy benefactor?* (Burrill Homes)
28 January 1998: *Historic York almshouse needs saviour says report* (Dorothy Wilson's)
23 January 1998: *Charity homes in on the needy* (Jane Wright's)
2 February 1998: *The York buildings left behind by charity* (Feature article)
7 October 1998: *Future looks brighter for old building: almshouses to get a facelift* (Dorothy Wilson's)
24 February 1999: *History in the making for city homes* (Dorothy Wilson's)
12 April 1999: *History lesson in a quiet corner of Skeldergate* (Anne Middleton's. Review of *A quiet corner of Skeldergate* by Andrea Yarnell Dakin)
21 May 2004: *Almshouses on parade* (Wandesford House upgrade)
7 March 2007: *Call to almshouses* (visit of Lord Mayor to Wandesford House)

Internet sources

Almshouse Association. http://www.almshouses.info/
Archaeological Planning Consultancy [reports] http://www.archaeologicalplanningconsultancy.co.uk/index.php
Archaeology of York : Web series 2.
http://www.iadb.co.uk/dccook/discussion.php
Charity Commission. http://www.charity-commission.gov.uk/
Sector study (Housing Corporation), 41 (2004).
http://www.housingcorp.gov.uk/upload/pdf/SS_41.pdf
Office for National Statistics. http://www.statistics.gov.uk
Woods, Richard: "The Sunday Times rich list 2010: fortunes of super rich soar by a third" in *Times online* (April 25 2010).
http://business.timesonline.co.uk/tol/business/specials/rich_list/article7107299.ece

Map of York (1: 1056, 5ft to 1 statute mile) by Captain Tucker RE, for the Ordnance Map Office; published in 21 sheets Sept-Oct 1852 (Held at York City Archives)
Online version by City of York Council and Beacon Dodsworth.
http://york1852.org/

Literary references

Apocrypha: Book of Tobit
Bible: Old Testament: Isaiah; New Testament: esp. the Gospels, and St Paul's First letter to the Corinthians.
Chaucer, Geoffrey *The prologue to the Canterbury tales*
Clare, John *The parish: a satire*
Copland, Robert *The hye way to the spytal house* (W.C. Hazlitt, *Remains of the early popular poetry of England, iv.* 1866)
Dickens, Charles *Bleak House; Oliver Twist; Our mutual friend; Sketches by Boz*; "A walk in a workhouse" in *Household words* 25 May 1850
Hardy, Thomas *Far from the madding crowd*
Kingsley, Charles *The water babies*
Nashe, Thomas *The unfortunate traveller*
Pearson, Hesketh *The Smith of Smiths*
Sandford, Jeremy *Cathy come home* (BBC1 Wednesday Play, directed by Ken Loach, broadcast 16 November 1966)
Trollope, Anthony *The warden*

Note

The manuscript notes belonging to W.K. Sessions will, post-publication, be passed to the Borthwick Institute at York University, together with other reference materials.

OLD PEOPLE

Report of a Survey Committee

on the problems of ageing

and the care of old people

under the chairmanship

of

B. SEEBOHM ROWNTREE
C.H., LL.D.

Published for the trustees of
THE NUFFIELD FOUNDATION
by Geoffrey Cumberlege, Oxford University Press, London · *1947*

APPENDIX 13

Report on the Almshouses of York

by W.K. Sessions: 1947

SOME details of the value of the almshouse charities of York were given in Appendix 2. The following report gives supplementary information.[1] There are altogether twenty groups of almshouses, the majority of which are described as hospitals, and they provide accommodation for 177 old people. The names, incomes, capital value, and accommodation provided in these twenty-two groups of almshouses are given below:

	Type of charity	*Maximum no. of residents*	*Income (approx.)*	*Value of buildings (approx.)*
			£	£
Thomas Fothergill Homes	Municipal	10	409	5,000
Ingram's Hospital	Municipal	11	88	2,500
Middleton's Hospital	Municipal	12	132	4,500
St. Catherine's Hospital	Municipal	4	85	1,625
St. Thomas's Hospital	Municipal	12	94	3,075
Terry Memorial Homes	Municipal	4	6	1,000
Sir H. Thomson's Hospital	Municipal	6	187	To be demolished
Sir R. Watter's Hospital	Municipal	7	32	1,500
Sister Wilson's Home	Municipal	10	120	1,200
Barstow's Hospital	Independent	4	57	1,100
Burrill Homes	Independent	7	290	10,250
Colton's Hospital	Independent	12	100	3,000
Harrison's Hospital	Independent	8	250	1,400
Lady Hewley's Hospital	Independent	10	340	3,121
Mason's Hospital	Independent	5	45	1,375
Cameron Walker Homes	Independent	11	256	2,400
Wandesford's Hospital	Independent	10	350	3,750
Dorothy Wilson's Hospital	Independent	16	..	6,000
Ellen Wilson's Hospital	Parochial	6	284	930
Winterscale's Hospital	Parochial	6	209	To be demolished

Note. Site-value is estimated to average £400 for each of the 20 groups of almshouses.

In the various groups of almshouses, there are at present (June 1946) approximately twenty vacancies, although less than half of them will be filled as the others are in almshouses that are to be demolished. The only accommodation in the almshouses for old men is:

Thomas Fothergill Homes	5
Terry Memorial Homes	2 (with their wives)
Sir H. Thomson's Hospital (about to be demolished)	6
Cameron Walker Homes	2 (with their wives)

All the remaining accommodation is for solitary old women.

Tables showing in detail the amenities provided in the almshouses are given on pp. 184–7, but it is intended first to give a brief description and a photograph of each.

1. *Thomas Fothergill Homes*

Thomas Fothergill Homes consist of ten ground-floor dwellings erected in 1935, half for men and half for women. They are attractively built of brick and tile in an L-shape. Each dwelling contains an entrance-hall, bedroom, living-room, kitchen with hot and cold water, bathroom, and water-closet. Each resident receives a pension of 10*s.* a week.

2. *Ingram's Hospital*

Ingram's Hospital comprises eleven dwellings in a row, built in 1640. The dwellings are of two stories, except for the centre one (occupied by a caretaker), which has an extra story. Each dwelling has a living-room with a large open fireplace; a narrow winding staircase leads to an unheated bedroom. Each resident has a wash-house, with a sink and cold tap; there is no bath. There are four outside water-closets. A pension of 2*s.* a week is paid to each resident. Radical structural alteration would be necessary to provide a good standard of amenity.

3. *Middleton's Hospital*

Middleton's Hospital, built in 1829 and modernized in 1939, stands back from a street with many warehouses. It consists of a substantial two-storied building; each of the eleven residents has two rooms and a sink with a cold tap. There is a water-closet on each floor, and one bathroom with a gas-heater. One room is not used owing to damp. The residents receive 2*s.* a week each and small additional sums from another charity.

4. *St. Catherine's Hospital*

St. Catherine's Hospital is a small, two-storied building of weathered brick and with a tiled roof. It was built in 1850, and is divided into four dwellings—two on each floor. Each dwelling has a bedroom and a living-room. There is a sink with cold tap on each floor, and there are two water-closets in the back-yard. There is neither a garden nor a back-door; for lack of a back-door (which could be remedied at small cost) coal and refuse have to be carried through the house. One of the dwellings on the ground floor is damp. Each resident receives a weekly pension of 7*s.*

5. *St. Thomas's Hospital*

St. Thomas's Hospital is built on a corner site, with the city walls immediately behind it, as can be seen in the photograph. A wall unfortunately prevents access from the small back-yard to the grass slope which fringes the city walls. The Hospital was built in 1862; it is a large building in yellow brick with a slate roof. The garden, which is at the side of the house, has been leased as an allotment, so that there is now only the very narrow frontal strip. There are six bed-sitting rooms on each of the two floors; the passages are dark and the staircase is of stone. The rooms have an open range and cupboards for food and

1. THOMAS FOTHERGILL HOMES

2. INGRAM'S HOSPITAL

3. MIDDLETON'S HOSPITAL

4. ST. CATHERINE'S HOSPITAL

5. ST. THOMAS'S HOSPITAL

6. TERRY MEMORIAL HOMES

7. SIR HENRY THOMSON'S HOSPITAL

8. SIR ROBERT WATTER'S HOSPITAL

coal. Four rooms have sinks with cold taps and the remaining eight rooms share four sinks. There are two water-closets, one on each floor. Residents receive a pension of 2*s*. 3*d*. a week and also a small grant from another charity. A greater degree of comfort could be provided in this hospital at moderate cost.

6. *Terry Memorial Homes*

Terry Memorial Homes were erected in 1899 in what was then part of the front garden of Middleton's Hospital. They are built of brick and tiles and consist of ground-floor dwellings for two aged couples. Each dwelling consists of a living-room with bed-recess, a scullery with sink and cold tap, and a water-closet under the back porch. One of the male residents looks after the grounds of both Middleton's Hospital and the Terry Homes. No pensions are paid to residents.

7. *Sir Henry Thomson's Hospital*

Sir Henry Thomson's Hospital is a two-story building of brick and tile, erected in 1700. Accommodation was provided for six men, each of whom had a living-room and a bedroom. There are two water-closets in the back-yard, and one outside water-tap. The building is subject to a demolition order; there is only one resident, and the vacancies are not being filled. The resident receives a weekly pension of 15*s*.

8. *Sir Robert Watter's Hospital*

Sir Robert Watter's Hospital is a two-story building in brick with a tiled roof, erected in 1844. There are seven dwellings in all; three have their doors on the street and three on the back-yard. Each of these six dwellings consists of a living-room on the ground floor and a bedroom above, reached by a steep and narrow staircase. The seventh dwelling consists of two rooms on the first floor reached by a flight of stairs leading off a dark covered passage running from front to back. There are two outside water-closets and only one tap, which also is in the back-yard. The ground-floor rooms suffer from black-damp and there is probably no damp-course. No pensions are paid but there are certain benefits in kind. It is doubtful whether this building could be made more healthy or more comfortable by modernization; complete rebuilding may be necessary.

9. *Sister Wilson's Home*

Sister Wilson's Home is a two-story building of brick and slate, erected in 1885. In addition to the building illustrated, there is a second building at the bottom of the garden facing a back-lane. The accommodation for individual residents varies a good deal, from a single room to a living-room and two bedrooms. Water-closets are shared and are all in the garden. One house has an indoor sink for each dwelling, but for the other house the sinks and taps are in an outside wash-house and are shared. Three rooms are said to have been empty 'for a long time' and there is dampness in some dwellings. A pension of 4*s*. a week is paid to each resident. The amenities and general comfort of these dwellings could probably be a good deal improved at a moderate cost.

10. *Barstow's Hospital*

Barstow's Hospital is of brick and slate, and was built in 1859. It consists of four bed-sitting rooms, all on the ground floor. There is one indoor water-closet and a communal sink and cold tap in the passage. Each resident receives 3*s*. 6*d*. a week and some benefits in kind.

11. *Burrill Homes*

Burrill Homes were built in 1931, and are of brick and tiles; they consist of seven ground-floor dwellings, each containing a bedroom, sitting-room, kitchen with hot and cold water, pantry, bathroom, and water-closet. The residents receive a pension of 10*s*. a week each and free light and water. A part-time gardener is employed.

12. *Colton's Hospital*

Colton's Hospital is built in a square, and consists of twelve ground-floor dwellings of brick and tiles. Eight were built in 1910 and four some twenty years later. Each dwelling has a living-room, bedroom, a pantry with sink and cold tap, and an inside water-closet. Residents receive a pension of 12*s*. 6*d*. a month.

13. *Harrison's Hospital*

Harrison's Hospital consists of eight lofty single-storied buildings of grey stone and slate, standing in attractive but unkept grounds. In the centre of the row of dwellings is a disused chapel, which is in poor repair both inside and out. The outside of the dwellings likewise needs a good deal of repair. Each has a bed-sitting room, separate kitchen with a sink and cold tap, and an outside water-closet. Residents receive a pension of 10*s*. weekly, but there is no fund available for structural repairs, for which approximately £500 is needed.

14. *Lady Hewley's Hospital*

Lady Hewley's Hospital consists of a block of nine stone dwellings running at right-angles to the road. In addition there is a chapel, which is in regular use, and a house for a caretaker. Each dwelling has a sitting-room and a kitchen, with sink and cold tap, on the ground floor, and an attractive bedroom and a boxroom on the first floor. There is an outside water-closet for each dwelling. Pensions of 10*s*. or 15*s*. are paid to residents according to their other resources.

15. *Mason's Hospital*

Mason's hospital is a three-storied brick building, one room wide, in a narrow central street of shops. There are five bed-sitting rooms, each with a sink and cold tap, and a communal wash-room. There is a water-closet in the back-yard. The roof leaks and is in urgent need of repair; the general decorative condition of the property is poor. A small pension, averaging about £7 a year, is paid to residents, but there is no adequate fund for maintenance of the building.

9. SISTER WILSON'S HOME

10. BARSTOW'S HOSPITAL

11. BURRILL HOMES

12. COLTON'S HOSPITAL

13. HARRISON'S HOSPITAL

14. LADY HEWLEY'S HOSPITAL

15. MASON'S HOSPITAL

16. CAMERON WALKER HOMES

17. WANDESFORD'S HOSPITAL

16. *Cameron Walker Homes*

Cameron Walker Homes consist of twelve dwellings in a two-storied building of stone with a slate roof, erected in 1912. The occupants are ten old women, and two elderly married couples who are the caretakers and the former caretakers, the latter couple being now too old to work. Each dwelling contains a bedroom, a living-room, and a small scullery with sink and cold tap. There are six inside water-closets and a well-fitted communal wash-house. The residents receive a pension of 5*s*. 6*d*. a week each and free light and water.

17. *Wandesford's Hospital*

Wandesford's Hospital was built in 1739, of brick with a tiled roof, and was re-roofed some twenty years ago. There are ten residents and a female caretaker. Each resident has a living-room and kitchenette on the ground floor and a bedroom on the first floor. There are two indoor water-closets and one bathroom. Three dwellings have sinks; the inhabitants of the other eight dwellings use communal sinks and taps in the passages. A chapel on the first floor is no longer in regular use. Payments and gifts in kind are made to the residents to an average of nearly £20 a head per annum.

18. *Dorothy Wilson's Hospital*

Dorothy Wilson's Hospital is a large three-storied building of red brick and tiles, erected in 1812. There is a caretaker's house behind the main building. The three floors contain sixteen bed-sitting rooms; there are a water-closet and a sink with a cold tap on each floor. The residents receive approximately £200 a year between them, in amounts varying according to their other resources.

19. *Ellen Wilson's Hospital*

Ellen Wilson's Hospital consists of a row of six single-storied dwellings, built in 1894 of brick with a tiled roof, and placed at right-angles to the street. Each dwelling contains a living-room with bed-recess, a small kitchen with sink and cold tap, and an indoor water-closet. Residents receive a weekly pension of 8*s*.

20. *Winterscale's Hospital*

Winterscale's Hospital, built in 1723, consists of two-storied buildings around the courtyard shown in the illustration. The ground floor is divided into six bed-sitting rooms, and above there is a long gallery and a malt-kiln; the latter is now disused. Each bed-sitting room has a sink and cold tap, and there are three outside water-closets. The rooms, only three of which are occupied, are wet, and the whole property is in poor repair and is due for demolition. Residents receive a weekly pension of 7*s*. 3*d*.

21. *Merchant Taylors' Hospital*

Merchant Taylors' Hospital consists of a row of four almshouses, built in 1730 of brick with tiled roof, and standing in the grounds of the

Merchant Taylors' Hall. They are now used to house the caretaker of the Hall, and there are no longer any aged residents.

22. *St. Crux Almshouses*

Although mentioned in official records early in the present century, this almshouse cannot now be traced and it is probably no longer in existence.

STANDARDS OF ACCOMMODATION

The following tables summarize in convenient form the standards of accommodation of the various almshouses:

TABLE I

	Date of building	*Present residents*		*Present vacancies*		Caretaker
		M.	*F.*	*M.*	*F.*	
A. *Municipal*						
1. Fothergill	1935	5	5	..	..	Yes for women
2. Ingram	1640	..	11	..	..	Yes
3. Middleton	1829	..	11	..	..	Yes
4. St. Catherine	1850	..	4	..	..	No
5. St. Thomas	1862	..	10	..	2	No
6. Terry	1899	2	2	..	..	No
7. Thomson	1700	1	..	5*	..	No
8. Watter	1844	..	7	..	..	No
9. Sister Wilson	1885	..	6	..	3	No
		8	56	5	5	
		(Total 74)				
B. *Independent*						
10. Barstow	1859	..	4	..	..	No
11. Burrill	1931	..	7	..	..	Yes
12. Colton	1910 and 1930	..	12	..	..	No
13. Harrison	1846	..	6	..	2	No
14. Hewley	1839	..	10	..	..	Yes
15. Mason	*c.* 1750	..	4	..	1	No
16. Walker	1912	2	12	..	..	Yes: a couple
17. Wandesford	1739	..	10	..	1	Yes
18. D. Wilson	1812	..	17	..	..	Yes
		2	82	..	4	
		(Total 88)				
C. *Parochial*						
19. E. Wilson	1894	..	6	..	..	No
20. Winterscale	1723	..	3	..	3*	No
		..	9	..	3	
		(Total 12)				
Grand total	..	10	147	5	12	
		(Total 174)				

Note. Vacancies marked * are not being filled, as the buildings concerned are due for demolition.

18. DOROTHY WILSON'S HOSPITAL

19. ELLEN WILSON'S HOSPITAL

20. WINTERSCALE'S HOSPITAL

21. MERCHANT TAYLORS' HOSPITAL

TABLE II

	Selection restrictions	*One room per pers.*	*More than one room*	*Stairs*		*Garden*		
				Yes	*No*	*Yes*	*No*	*Hardly any*
A. *Municipal*								
1. Fothergill	York	..	10	..	10	10	..	..
2. Ingram	York widows	..	11	11	..	11	..	..
3. Middleton	Freemen's widows	..	11	6	5	11	..	..
4. St. Catherine	York	..	4	2	2	..	4	..
5. St. Thomas	York	12	..	6	6	..	..	12
6. Terry	York	4	..	..	4	4	..	..
7. Thomson	York men	..	6	6	..	..	6	..
8. Watter	York	..	7	7	..	..	..	7
9. Sister Wilson	York	6	3	6	3	9	..	..
		22	52	44	30	45	10	19
		(Total 74)		(Total 74)		(Total 74)		
B. *Independent*								
10. Barstow	York	4	..	..	4	..	4	..
11. Burrill	York	..	7	1	6	7	..	..
12. Colton	Pref. Unitarians	..	12	..	12	12	..	..
13. Harrison	C. of E.	..	8	..	8	8	..	..
14. Hewley	Not only York: Congreg. Presbyt. and Baptist	..	10	10	..	10	..	..
15. Mason	Not only York	5	..	4	1	..	5	..
16. Walker	York or near	..	14	6	8	14	..	..
17. Wandesford	Spinsters, C. of E.	..	11	11	..	11	..	..
18. D. Wilson	Women in domestic service in York	16	1	12	5	..	..	17
		25	63	44	44	62	9	17
		(Total 88)		(Total 88)		(Total 88)		
C. *Parochial*								
19. E. Wilson	P. of St. Lawrence	..	6	..	6	6	..	..
20. Winterscale	P. of St. Margaret	6	..	..	6	..	6	..
		6	6	0	12	6	6	0
		(Total 12)		(Total 12)		(Total 12)		
Grand total	..	53	121	88	86	113	25	36
		(Total 174)		(Total 174)		(Total 174)		

TABLE III

	Lighting		*Baths*			*Running water*		
	Elect.	*Gas*	*Indiv.*	*Share one*	*None*	*Sink in own rooms*	*Share in-doors*	*Out-side tap*
A. *Municipal*								
1. Fothergill	10	..	10	..	..	10	..	..
2. Ingram	..	11	..	..	11	..	..	11
3. Middleton	11	..	..	11	..	11	..	..
4. St. Catherine	..	4	..	..	4	..	4	..
5. St. Thomas	..	12	..	..	12	4	8	..
6. Terry	..	4	..	..	4	..	4*	..
7. Thomson	..	6	..	..	6	..	..	6
8. Watter	..	7	..	..	7	..	..	7
9. Sister Wilson	..	9	..	..	9	4	..	5
	21	53	10	11	53	29	16	29
	(Total 74)		(Total 74)			(Total 74)		
B. *Independent*								
10. Barstow	..	4	..	..	4	..	4	..
11. Burrill	7	..	7	..	..	7	..	..
12. Colton	..	12	..	..	12	12	..	..
13. Harrison	..	8	..	..	8	8	..	..
14. Hewley	10	..	..	..	10	10	..	..
15. Mason	..	5	..	..	5	5	..	..
16. Walker	14	..	..	..	14	10	4*	..
17. Wandesford	11	..	..	11	..	3	8	..
18. D. Wilson	17	..	..	..	17	1	16	..
	59	29	7	11	70	56	32	0
	(Total 88)		(Total 88)			(Total 88)		
C. *Parochial*								
19. E. Wilson	6	..	..	..	6	6	..	..
20. Winterscale	..	6	..	..	6	6	..	..
	6	6	0	0	12	12	0	0
	(Total 12)		(Total 12)			(Total 12)		
Grand total	86	88	17	22	135	97	48	29
	(Total 174)		(Total 174)			(Total 174)		

Note. * denotes sharing by married couples.

TABLE IV

	Water-closets				Coal-stores	
	Indoor		Outdoor			
	Indiv.	Share	Indiv.	Share	Satis.	Unsatis.
A. *Municipal*						
1. Fothergill	10	..	..	..	10	..
2. Ingram	..	..	..	11	6	5
3. Middleton	..	11	..	..	..	11
4. St. Catherine	..	..	..	4	..	4
5. St. Thomas	..	12	..	..	..	12
6. Terry	..	4*	..	..	4	..
7. Thomson	..	..	..	6	..	6
8. Watter	..	..	..	7	..	7
9. S. Wilson	..	..	..	9	..	9
	10	27	0	37	20	54
	(Total 74)				(Total 74)	
B. *Independent*						
10. Barstow	..	4	..	..	..	4
11. Burrill	7	..	..	..	7	..
12. Colton	12	..	..	..	12	..
13. Harrison	..	..	8	..	..	8
14. Hewley	..	..	10	..	10	..
15. Mason	..	..	..	5	5	..
16. Walker	..	14	..	..	9	5
17. Wandesford	..	11	..	..	..	11
18. D. Wilson	1	16	..	..	12	5
	20	45	18	5	55	33
	(Total 88)				(Total 88)	
C. *Parochial*						
19. E. Wilson	6	..	..	..	..	6
20. Winterscale	..	..	..	6	..	6
	6	0	0	6	0	12
	(Total 12)				(Total 12)	
Grand total	36	72	18	48	75	99
	(Total 174)				(Total 174)	

Note. * denotes sharing by married couples.

Comments on the Tables

(*a*) *Date of building.* It will be noted that 5 of the 'hospitals' were built before the nineteenth century; 5 in the first half of the nineteenth century; 6 in the second half of the nineteenth century; and 4 in the present century.

(*b*) *Residents.* The total accommodation for 159 women and only 15 men would seem out of proportion, despite the extra longevity of women and the greater willingness of men to go to a Public Assistance

Institution. Unless Thomson's Hospital is rebuilt the accommodation for men will be reduced to 9, and only 5 of these places are for bachelors or widowers.

Eight out of the 9 available vacancies have been unfilled for long periods, extending into years in some instances. In at least one of the trust deeds concerned it is expressly laid down that vacancies shall be filled within a period of 3 months.

(*c*) *Caretaker.* Eight out of the 20 almshouse-groups have a caretaker, who is usually a somewhat younger woman of late middle-age. The Cameron Walker Homes have a recently retired couple as caretakers, and this could with advantage be the case at the Fothergill Homes, where the caretaker is a woman although half the residents are men.

(*d*) *Accommodation.* Fifty-three out of 174 (or less than one-third) have only one room, while half the total number have stairs to climb. Nevertheless, the fact that there is ground-floor accommodation for half the number of residents is very satisfactory compared with, say, the average position in Homes for the Aged. One hundred and thirteen have adequate gardens, which are often very attractive; 61 residents have no garden space or hardly any.

(*e*) *Lighting.* Half the residents have only gas lighting, and in this respect the independent 'hospitals' are considerably ahead of the municipal ones. More than two-thirds of the municipal 'hospitals' still have gas lighting (53 out of 74); while only one-third of the independent 'hospitals' are lit by gas (29 out of 88). In either case there is considerable room for improvement, for gas is a danger to old people with stiffening fingers. Apart from lighting, gas is used in a very large proportion of the dwellings in the form of gas pokers, rings, and small or large cookers. Many of the fitments are primitive—for instance, portable gas rings connected by rubber nozzles to ancient gas-brackets on the wall. On general grounds it is desirable for electricity to be installed in all the 'hospitals', not only for lighting but also to allow the provision of electric kettles and breakfast cookers in all the rooms.

(*f*) *Wireless sets.* Installation of electricity, in addition to giving safer lighting and other facilities, would assist more of the residents to have wireless sets. Maintaining a battery-set is too complex for most old people, and the radio relay service is beyond the pockets of many.

(*g*) *Baths.* Only 17 residents have individual baths, and a further 22 people share 2 baths. The remaining 135 have no bath, no washbasin, and no access to running hot water; they have to heat by kettle or pan for all needs. One hears the opinion expressed that many of these folk have never been in the habit of taking baths and that therefore there is no need to provide them. Several of the almspeople in the survey deplored, however, the absence of a bath, and described the hard work involved in taking a bath in a zinc tub for which all the water has to be carried. Experience gained in Homes for old people shows both the importance of encouraging regular baths in order to keep dirt and disease to a minimum, and the improvement in health that results. Shortage of space would make it difficult to install baths at many of the 'hospitals', but there are a number of instances where inquiries could usefully be made of an architect whether space could be arranged for a sitz-bath—a type particularly suitable for elderly folk.

(*h*) *Running cold water.* Ninety-seven residents have sinks with cold-water taps in their own quarters; a further 48 share indoor taps; and 29 have to go out of doors to a tap. Here again the independent 'hospitals' are more advanced than the municipal ones: all the 29 residents who have only outside taps live in 'hospitals' run by the municipal trustees, and there are no outside taps in the independent 'hospitals'.

(*j*) *Water-closets.* Only 36 residents have individual indoor water-closets, and a further 72 share indoor ones. Sixty-six old people have to go out of doors to a closet, 18 of these having individual ones. Once again the municipal 'hospitals' have a noticeably bigger proportion of people who have the lowest standard of amenity: 37 residents in 5 of the municipal 'hospitals' (one of them due for demolition) have to share outside water-closets, as compared with 11 residents in 2 of the other 'hospitals' (one of them due for demolition).

(*k*) *Coal storage.* Shovelling coal and clearing fires is heavy labour for many old people, and yet an open fire is a far greater comfort to the big majority than, say, a gas or electric fire. In Table 4 coal storage has been assessed as unsatisfactory (*a*) if the old person has to go out of doors for it; (*b*) if the coal is kept in a cupboard in the living-room; or (*c*) if the coal is for other reasons not easily accessible. By these standards 54 out of the 74 residents in the municipal 'hospitals' have unsatisfactory coal-storage arrangements, as compared with 45 out of 100 in the other 'hospitals'.

General observations

(*a*) *Administration.* Eighteen out of the 20 'hospitals' have solicitors as clerk or secretary to the trustees. Of the other two, Dorothy Wilson's is administered by a firm of estate-agents, and Lady Hewley's is managed by sub-trustees of the main Lady Hewley Trust in Liverpool. Generally the secretaries are responsible for the running of the 'hospitals'.

(*b*) *Relative standards.* Details have been given in the previous section showing that the average standard of amenity of the municipal 'hospitals' is in several important respects below that of the independent 'hospitals'. One might have expected the reverse to be the case, in view of the Charity Commission's recommendations on various occasions in the past concerning the amalgamation of endowed charities. While co-ordination of the various charities may be desirable to prevent overlapping and to secure an equitable sharing of available benefits, it would nevertheless appear that having a separate group of people responsible for each individual 'hospital' may result in improved conditions for the residents. There might well be advantages in appointing a small house-committee of people living in the immediate neighbourhood for each of the municipal 'hospitals'.

(*c*) *Demolition.* The following hospitals are probably, owing to age or damp, no longer adequate for habitation:

(1) Municipal . . Sir Henry Thomson's, built 1700.
Sir Robert Watter's, built 1844, but now very damp.

(2) Independent . Mason's, built about 1750: dark, and on three floors.

(3) Parochial . . Winterscale's, built 1723.

Of these, Sir Henry Thomson's and Winterscale's are already due for demolition, and vacancies in them are not being filled. The view of the Committee has already been expressed in paragraph 71 of this report (p. 30) on how funds available after demolition can most suitably be used. There are two other 'hospitals' which were built more than 200 years ago. Wandesford's in Bootham was erected in 1739 and is still providing comfortable accommodation. The oldest of them all, Ingram's in Bootham, was built in 1640 and has an attractive frontage on the street, which ought to be preserved; yet it is difficult to see how the awkward conditions of living could be much improved without very substantial structural alteration of the back parts of the building, including chimneys, stairways, and outbuildings. When the housing situation becomes easier, Ingram's might well be closed as a 'hospital' and retained as an architectural monument.

(*d*) *Liaison and welfare.* The many preoccupations of the war-years have probably caused some diminution in the frequency with which the trustees and secretaries of some of the 'hospitals' have been able to visit the residents. In two 'hospitals', where a cheerful and independent spirit prevailed, the residents were glad to be left largely to themselves; but in others regular visiting by interested people would be welcomed.

(*e*) *Ill health.* An appreciable number of the almshouse residents were finding it difficult to look after themselves adequately, owing to increasing feebleness. In a few instances a relative was able to visit daily in order to clear the fire, bring food-supplies and a cooked meal, tidy the room, and assist with washing, dressing, and bed-making; but for others, whose infirmities were growing, the situation was considerably less satisfactory, and there was frequently a deep-rooted fear of being moved to the Public Assistance Institution's infirmary. Furthermore, the elderly widowers found it especially difficult to fend entirely for themselves.

Index

Bold denotes illustration